Economic Geography

This book series serves as a broad platform for scientific contributions in the field of Economic Geography and its sub-disciplines. *Economic Geography* wants to explore theoretical approaches and new perspectives and developments in the field of contemporary economic geography. The series welcomes proposals on the geography of economic systems and spaces, geographies of transnational investments and trade, globalization, urban economic geography, development geography, climate and environmental economic geography and other forms of spatial organization and distribution of economic activities or assets.

Some topics covered by the series are:

- Geography of innovation, knowledge and learning
- Geographies of retailing and consumption spaces
- Geographies of finance and money
- Neoliberal transformation, urban poverty and labor geography
- Value chain and global production networks
- Agro-food systems and food geographies
- Globalization, crisis and regional inequalities
- Regional growth and competitiveness
- Social and human capital, regional entrepreneurship
- Local and regional economic development, practice and policy
- New service economy and changing economic structures of metropolitan city regions
- Industrial clustering and agglomeration economies in manufacturing industry
- Geography of resources and goods
- Leisure and tourism geography

Publishing a broad portfolio of peer-reviewed scientific books *Economic Geography* contains research monographs, edited volumes, advanced and undergraduate level textbooks, as well as conference proceedings. The books can range from theoretical approaches to empirical studies and contain interdisciplinary approaches, case studies and best-practice assessments. Comparative studies between regions of all spatial scales are also welcome in this series. Economic Geography appeals to scientists, practitioners and students in the field.
If you are interested in contributing to this book series, please contact the Publisher.

More information about this series at http://www.springer.com/series/15653

Fabio Betioli Contel

The Financialization of the Brazilian Territory

From Global Forces to Local Dynamisms

Springer

Fabio Betioli Contel
Department of Geography
University of São Paulo
São Paulo, Brazil

ISSN 2520-1417 ISSN 2520-1425 (electronic)
Economic Geography
ISBN 978-3-030-40292-1 ISBN 978-3-030-40293-8 (eBook)
https://doi.org/10.1007/978-3-030-40293-8

This Springer imprint is published by the registered company Springer Nature Switzerland AG
The registered company address is: Gewerbestrasse 11, 6330 Cham, Switzerland

It is finance capital that appears to be the all-pervading form of capital, that form which, like nature, suffers from a horror vaccui , *since it rushes to fill every* "vacuum", *whether in a 'tropical', 'sub-tropical', or 'polar' region, if only profits flow in sufficient quantities*

Nicolai Bukharin in Imperialism and the world economy *(1915, 58)*

Political economy cannot be done without given space. Space can be defined as the result of a permanent interaction between, on the one hand, the accumulated labor, in the form of infrastructures and machines that overlap with nature and, on the other hand, the present labor, distributed on these forms from past. Dead labor, on which living labor is exercised [...] constitute exactly the geographical space

Milton Santos in Por uma economia política da cidade *(1994, 115)*

Acknowledgements

This book is fundamentally the result of a sabbatical year held between 2017 and 2018 at the School of Geography at University of Oxford. Without this *séjour* when I could devote full time to reading, searching, and working with bibliographical and empirical materials, it would never have been possible to achieve this result. First of all, I would like to thank the generous reception of my colleague and friend Darek Wójcik. The partnership and the learning made during this period were fundamental for the advancement of the book and to my own academic formation. Over these 12 months of residence in England it was possible to participate in seminars organized by the FINGEO network which were also primordial for the maturation of several of the proposals presented here. In addition to all FINGEO colleagues who I am pleased to be contacting more often now, I would like to thank Martin Sokol and David Bassens for their friendship and fruitful academic cooperation.

This sabbatical year would also not have been possible without the combination of a series of supports received from different Brazilian public academic institutions. It was fundamental, firstly, the permission of the Department of Geography and the Graduate Program in Human Geography of the University of São Paulo for me to have this 1-year leave, institutions that I am deeply grateful for. The financial assistance of the Graduate Program for the translation of this book into English was also fundamental (all parts quoted in languages other than English are from our translation). No less important were the scholarships received from the São Paulo Research Foundation (FAPESP) and the National Council for Scientific and Technological Development (CNPq), without which it would not have been possible to carry out this work and research abroad. I would also like to thank these two institutions, which continue to be essential for the maintenance of the public and autonomous nature of researches conducted in Brazilian universities.

I should also like to say a sincere word of thanks to Juliana Pitanguy, Publishing Editor of Springer, who from the beginning of our first conversations in Boston (2016) has always been extremely professional and encouraging, but generous as well.

Also to be fair with my inescapable Brazilian and Latin American roots, I would like to make a special mention to my friends Mónica Arroyo, Ricardo Mendes, and María Laura Silveira for their warm and genuine friendship since the remote times of our formation as researchers in the Laboratory of Territorial Planning (LABOPLAN) of the Department of Geography of the University of São Paulo. Thanks also to my undergraduate and graduate students, who are performing an important work in Brazil to understand the uses that financial firms and institutions make of geographical space. I thank symbolically to all my students through the memory of the talented Thiago Marques Guerim, who so suddenly and early left us.

Last but not least, I would like to thank my family, and specially my beloved wife Camille, for her warmth and understanding throughout the process of researching and writing this book. From the Oxford internship to the final writing period in São Paulo, Camille was extremely amorous and cooperative in all phases of this long endeavor, which escaped the timelines and planning devised at the beginning of the process.

Introduction

The financialization of the territory and the particularities of the Brazilian sociospatial formation.

Like all economies in periphery of the capitalist system, the financialization of Brazilian territory evolved both by internal processes and by the influence of an engine linked to the expanding international capitalism. At every cycle of world economy and world politics, peripheric countries have to adapt to new directions that come from the center of the system. This adaptation, however, is never made in an inexorable and direct form. Due to the proper existence of a set of infrastructures, demographic and technical densities, preexistent productive circuits, and specific legal and normative contents, each sociospatial formation reacts in a different way to these external forces, as shown by Santos (1977) in his classic article entitled "Society and space: social formation as theory and method." To this expansion of capitalist nexuses originated in dynamic centers of world system toward periphery countries, Harvey (1982) gave the broad denomination of "spatial fix."

In respect to the financialization of Brazilian territory, this external influence was always present and determined great part of the main processes that are in the basis of our economic and financial system. Since the impulses given by the first foreign mortgage banks that settled in the early twentieth century, until the current internationalization of national market capital, much of what occurred in Brazil results from its historical-structural condition of dependent country that made the Brazilian territory a "derived space" (*espaço derivado*) (Santos [1975] 1979). Although the control of banking system has not been internationalized along the twentieth century, foreign indebtedness—public and private, and all strong conditionalities that this indebtedness generated—was one of the determinant factors for the growth of the country's subordination to the international financial system and the consequent nation's underdevelopment (Baer 1986).

The term financialization is a neologism that gained great diffusion in contemporary social sciences, with significant emphasis on geography (Engelen 2008; Pike and Pollard 2010; Sokol 2013; Aalbers 2015; Christophers 2015). The idea that is behind the term dates back to the beginning of the twentieth century, when important Marxist intellectuals identified the increase of the power of banks and capital markets, in the context of the rise of a "financial capitalism." For the pioneer book of Hilferding ([1910] 1985, 27), "no understanding of current economic trends, no kind of economic or political science is possible without knowledge of the laws and function of financial capital." For him, it would be impossible to understand financial capitalism without analysis of elements that until now are in the center of "financialization" discussions: (1) the role of money and banks; (2) the function of credit; (3) the analysis of corporations and monopolies; (4) the analysis of stock exchanges and commodity exchanges (i.e., of capital markets); and finally (5) the analysis of financial crises and their causes. The other major reference work of Marxism at the same period is Vladimir Lenin, which better verified that we would be living in the beginning of twentieth century "the turning point of the old capitalism to the new, from the domination of capital in general to the domination of financial capital" (Lenin [1917] 1984, 610).

The debate about what currently is denominated as "financialization," therefore, is not new; what happened was an increase in the importance of finances in all the so-called "social instances," such as the economy itself, but also instances like politics, society, law, culture, and geographic space.

During the entire twentieth century, several economists were concerned with the analysis of financial variables and processes, list that would be impossible to exhaust due to the amplitude and diversity of authors that focused on the theme. One of the most important intellectuals that analyzed finances, Schumpeter ([1911] 1997, 81), showed that credit would have the essential function of "discarding means of production (already employed in some place) of the circular flow and allocate them in new combinations," generating the innovations that brake the "circular flow of economy" and install a "creative destruction." Keynes ([1936] 1964), on the other hand, shows the countercyclical role that the state debt can have in moments of crisis, by placing public debt securities for sale and financing new investments of great magnitude, which in turn increase demand for labor and generate opportunities for (full) employment and other investments from all types of firms.

The debate in the social sciences—and especially in economics—on the topic of financialization and financial dominance came to the fore in the 1990s and gained more definitive contours in the year 2005. In a seminal text, Chesnais (1997) shows that as from the 1970 decade, with the crises of the Fordist accumulation regime, we would have entered in a "regime of world accumulation with financial dominance," with expressive changes in the relationship of big industries with the financial system, and the rise of new financial operators, beyond banks. This is the case for the major pension funds, mutual funds, sovereign wealth funds, insurance firms, credit cards companies, which make the financial weight greater, in addition to insufflating the own process of "securitization" of economy (making capitalism even more focused for the remuneration of shareholders). For Chesnais (1997, 75), "these financial operators are a qualitative new type that have been by far, the main beneficiaries of 'financial globalization'." For Guttmann and Plihon (2008, 581), other two major exponents of the Regulation School, this new feature of contemporary capitalism brings a reduction of the power of commercial and/or productive agents and "dominance of financial motivations" in their agency.

In the Anglo-Saxon literature, one of the first books that use systematically the term is of Martin, *Financialization of Daily Life* (2002). Martin shows the broader consequences of financialization on the local scale, of the lives of individuals and households, stating that it brings a "hypercompetitiveness" to the more banal events and behaviors of social daily life: "The furious proliferation of money talk in the media permeates the home to a degree difficult to imagine only a few years ago" (Martin 2002, 37). Gerard Epstein, one of the main proposers of the term, argues that financialization is a process that would have emerged in the 1970s, jointly with the rise of a neoliberal era and the consequent deregulation of practically all national and international markets (securities, currencies, labor, land markets, etc.). The financialization ascends, therefore, in a context of relative reduction of State power in respect to large companies (and the global market mechanisms in general), with consequences for all nations in the world. In the definition of the author, "financialization means the increasing role of financial motives, financial markets, financial actors and financial institutions in the operational of the domestic and international economies" (Epstein 2005, 3). According to Epstein, it seems incontestable that there are more than robust empirical elements to prove that there has been a rise of finances as a key variable of the current historical period we are living. The volume of transactions in stock exchange more than tripled from 1989 to 2004; the international financial markets have been moving greater amounts (mainly regarding international trade); the profits of financial companies are also even larger—and in some cases incomparably greater—than the profit of non-financial companies (Epstein 2005, 4).

Krippner (2005), another important theorist of financialization, points that this process is made in a context of a transition from an industrial economy to an economy of services (or "post-industrial" economy). Particularly important in this new phase of world economy organization—and mainly in wealthier nations—are the professions associated to financial services, which stipulated to use the acronym FIRE (Finance, Insurance and Real Estate professions) to nominate them. In the organization plan of companies, this rise of financial services importance reflects also in a vigorous outsourcing process and subsidiaries formation that elevate the efficacy of productive chains, but maintains the process control under the command of oligopolies (Krippner 2005, 189/190).

In this sense, and strongly influenced by the work of Giovanni Arrighi, Krippner (2005, 181) defines financialization as a "pattern of accumulation in which profit making occurs increasingly through financial channels rather than through trade and commodity production." For the author, the different meanings for the financialization concept brought by the literature reveal also other central characteristics of the process:

1. "ascendancy of the 'shareholder value' as a mode of corporate governance";
2. "growing dominance of capital market financial systems over bank-based financial systems";

3. "increasing political and economic power of a particular class grouping: the rentier class";
4. "explosion of financial trading with a myriad of new financial instruments" (Krippner 2005, 175).

Another central contribution to this debate was made by Lapavitsas (2011). For him, financialization may be understood as a "systemic transformation of the capitalist economy" (Lapavitsas 2011, 619). The financialization represents therefore "a shift of the economy in the direction of the financial sector" (*op. cit.*, 621) and would have three main characteristics:

1. "Large corporations have acquired independent financial skills—they have become financialized";
2. "Banks have turned toward mediating transactions in open markets, thus earning fees, commissions, and trading profits. They have also turned toward individuals in terms of lending and handling financial assets";
3. "Third, workers have become increasingly involved with the financial system both with regard to borrowing and to holding financial assets. The retreat of public provision in housing, health, education, pensions and so on has facilitated the financialization of individual income, as have stagnant real wages. The result has been the extraction of bank profits through direct transfers of personal revenue, a process called financial expropriation" (Lapavitsas 2011, 623).

In Brazil, the pioneer of financialization debate is José Carlos Braga, who wrote in 1993 an article entitled "*The financialization of wealth: the financial macrostructure and the new dynamics of central capitalisms*." Strongly based on the work of Karl Marx and Hyman Minsky, the author shows that the financialization "is the general expression of contemporary forms of defining, managing, and realizing wealth in capitalism. By financial dominance we learn, even conceptually, the fact that all corporations—even those typically industrial [...]—have in their financial investments, of retained profits or cash, a central element in the process of global accumulation of wealth" (Braga 1993, 26).[1]

More recently, Paulani (2009, 34) shows that Brazil was "a character in the history of the financialization of capitalism since its inception," due to the fact that both the debt crisis of the 1980s and the financial liberalization of the 1990s are a constitutive part of the articulation of global financial capitalism with the internal dynamics of the national economy. For the author, "the entire institutional outline

[1]Braga introduced the subject of *financial dominance* as early as 1985, even before his French colleagues from the Regulation School (Michel Aglietta, Robert Boyer, François Chesnais, and Robert Guttmann, who called attention to this process in the 1990s). His doctoral thesis presented in 1985 (which was published only in 2000) is titled *Temporality of Wealth. A contribution to the Theory of Capitalist Dynamics*. In this original work, Braga (1985, 220) argues that "financial capital is structuring (and de-structuring) capitalism in this twentieth century, and imposes financial dominance and monopolization on it (a sharp form of opposition of private wealth versus social wealth), a complex plot between monopolies and the capitalist state, without coming to a regulated and exempt of crisis society, on the contrary, since the State is not an entity outside the financial capital system".

that frames the functioning of Brazilian capitalism today was built on the principle of responding promptly and precisely to the interests of financial wealth, in particular of creditors and external investors" (Paulani 2017, 30), and claims that "there is no way out without the reversal of financialization."

The debate on finances in geography was always shyer, during practically the entire twentieth century. Since the English classics about commercial geography in the beginning of the past century, to the time before the publication of the remarkable works of Labasse (*Les Capitaux et la Région* in 1955 and *L'Espace Financier* in 1974), little was said about the theme. The economic geography of descriptive and empirical bias made in the early decades of twentieth century was much more concerned with the study of economic activities that were easily discernible in the landscape—and consequently more readily visible and mappable—such as agriculture, transport, industry, and trade, as we can see in the classical works of Chisolm (1889), Jones and Darkenwald (1941), Allix (1960), Juglas (1950), among so many other important economic geographers in the period. Prior to the 1990s, in addition to some sparse articles published (including the important synthesis of Donald Kerr in 1965), there are only two works in Anglo-Saxon geography that systematically cared on the subject of finance: Richard Bennett, with his little remembered *The Geography of Public Finance* (1980) and David Harvey, in his *The Limits to Capital* (1982).

In the possession of this discussion about the process of financialization, the first observation made when analyzing the Brazilian territory as a whole is the following: financialization as it is consensually defined by the mainstream currents of this bibliography occurs until today in Brazil more faintly, especially if we think of these new and sophisticated financial instruments that characterize contemporary securitization and that are largely produced and marketed in the main financial centers of the world system. The concrete action—and localization—of these most modern financial institutions and instruments, which characterize this kind of financialization, are basically limited to a few national metropolises, mainly São Paulo, but also Brasília and Rio de Janeiro (as we can apprehend from the works of Mariana Fix (2007) and Daniel Sanfelici (2013)) and to some agricultural areas where there is also a more financially oriented crops (Goldfarb 2015). In this sense, what is going on in countries like Brazil—and in all countries on the periphery or semi-periphery of capitalism—is more a *low-intensive financialization*, less dependent on complex/sophisticated financial products and services that are common within the *high-intensive financialization* process of rich countries.

The financialization of the territory, in the sense we are going to use in this book, can be defined as the concrete process of spatial diffusion of financial variables, whether material or immaterial, more directly linked to the granting of credit, or to the "creditization of the territory," as proposed by Milton Santos (1993) to understand the Brazilian geographical modern history. This diffusion, in turn, depends on a set of other variables that can be considered as geographic, be they the banks or stock exchanges and their networks, but also more general infrastructures like information transmission networks, equipment and data processing centers, computers, and even the objects that are produced, purchased, or sold with financial instruments. In a more empirical sense, this definition of the concept allows us to identify, in a more long-term perspective, the action in the national space of the main agent of the national financial system: commercial/universal banks. From a more theoretical perspective, this setting permits us to treat in an inseparable way the actions of the financial agents—and the financial circulation produced by them—with the infrastructures, technical systems, and objects that allow these actions and flows to exist. This definition is largely inspired in the contributions of the Brazilian geographer Milton Santos. In several of his most "recent" works (1993, 1994a, 1996, 2000) and in the book *O Brasil: Território e Sociedade no Início do Século XXI*, written with María Laura Silveira (2001), he analyzed this contradictory modernization and new uses of territory that finance provides (Santos 1993; Santos and Silveira 2001).

In identifying the diffusion of agricultural credit within the Brazilian territory since the 1960 decade, Santos (1993, 42) proposed the expression "creditization of the territory," a neologism that defined the diffusion of banking branches, wages, and agricultural credit granted by banks in the areas where they were installed. This creditization process is one of those responsible for the increase of the "organic composition of space" (in allusion to Marx's original concept of "organic composition of capital") and is the more complete manifestation of the "need for advanced capital" than this modernization of the territory induces. As the author further notes, "this creditization of the territory, the dispersion of a highly productive production, the expansion of capitalism, the exacerbation of movement would not be possible without the informatization of the Brazilian space" (Santos 1993, 44), i.e., without the dissemination of information networks that allow the operation of the most modern firms (including financial firms).

In their most important synthesis work on contemporary Brazilian territory, Santos and Silveira (2001) identify the political economy of the Brazilian national space highlighting the existence of these more sophisticated infrastructures and modern technical networks, intensive in science and information (configuring what the authors call "technical-scientific-informational *milieu*"). In analyzing the Brazilian financial system, they use the concept of "financialization of society and territory" to refer to the diffusion of "financial instruments" in the national space—mainly deposits and credits—but also the distribution of financial institutions and banking networks—which they nominate "banking topology" (Santos and Silveira 2001, 185). Understanding the importance of banks, stock exchanges, banking automation, agricultural financing is fundamental to also identify in national space the areas that are little, or not served by financial agents, and therefore will perform more subordinated functions in the capitalist territorial division of labor. Finally, the authors give to the banking system the leading role in this financialization of the territory: "this expansion (of the financial instruments) was largely controlled by private banks, which spread through the territory because of the opportunities to finance production and circulation (processes) highly dependent on capital anticipation" (Santos and Silveira, *op cit*, 186).

These propositions about the "creditization" and the "financialization" of the territory are supported by the definition of geographic space that Milton Santos himself proposed in the 1990s, a definition that seems to explain properly the current historical period and which constitutes the central nucleus of our theoretical frame. For Santos, the geographic space consists of both the human actions and the technical objects that allow these actions to take place:

> Space is formed by an indivisible, solidary and also contradictory set of systems of objects and systems of actions, not considered in isolation, but as the unique framework in which history takes place. [...] Objects have no philosophical reality, i.e., they do not allow knowledge if we see them separated from the systems of actions. The systems of actions also do not happen without the systems of objects (Santos 1996, 51)

Considering the financialization as a function of actors, material infrastructures and the norms that regulate them, invites us to think about the financial entities indissociably to the geographic ones. This work is intended to be theoretically grounded on this basic assumption. It is by using this definition that it was possible to work on the diffusion of the first global information transmission networks, which were the transatlantic cables. Without them, financial agents would not have built their global ascendancy over other economic circuits (agricultural, industrial, commercial, service, etc.) and could not have centralized control over some of the national financial markets, constructing then their contemporary hegemony, by acting from a few number of large global financial centers. Without these information networks, it would not be possible to develop international banking networks, to mobilize financial instruments and even offshore financial centers, especially those in small countries—or islands—physically distant from the main “onshore” global financial centers. Themes like these are extensively treated through the next chapters.

Finances, therefore, start having greater importance, for both daily lives of families, for the action of firms, of States, as well as for the economic system as a totality. It is exactly in this context that the current book intends to make a geographic approach of the financialization process of Brazilian territory. As also shows most part of this literature about finances and financialization, even if these variables have gained a relative greater economic and social autonomy, they are still dependent on other instances that are the basis of the historical totalization process, including the geographic—or spatial—instance.

The Brazilian financial and banking system, according to most studies on the subject—made primarily by economists—have the following main characteristics: low financial depth (relation between financial wealth and gross domestic product); economically and territorially restricted capital markets, compared to the enormous power and extent of commercial banks; extremely modern and efficient financial infrastructure, both in its technical and regulatory aspects; exacerbated concentration of São Paulo’s financial power in the context of the national urban network; incapacity of the system to effectively include and act ethically regarding lower income populations. The present book seeks to give a geographical interpretation of these phenomena by identifying the main infrastructures, agents, processes, and stages of development of the Brazilian financial and banking system.

The first chapter of the book shows the broader genesis of this financialization process, explaining the global scale of finance reorganization that starts after the

Second World War. It is as from this time on that a new international institutionality is established (mainly with the Bretton Woods Agreement and creation of International Monetary Fund) and also a new international division of labor is consolidated, with expressive concentration of economic-financial power in USA and in Western Europe (and with an increasing importance of financial market-places of New York and London). It was as from this moment that new technical systems for information transmissions were disseminated globally, primarily modern submarine cables, and then optical fiber networks and artificial satellites, that gives even more power and organizational strength to the banks and financial institutions to act in global scale. At this same period, three other fundamental processes occurred: the creation of offshore financial jurisdictions—the "turntables" of the international financial system, according to Labasse (1974)—that allowed a greater dynamism in the circulation of finance on a global scale; the process of indebtment of peripheral countries that became more subordinated actors in this international division of labor with a "financial dominance"; and last but not least, the global spread of neoliberal ideology, with serious consequences for the concentration of wealth and the increase of poverty in peripheral countries such as Brazil. Within this new historical and geographical context, in the last part of Chap. 1, we define the relative position of the city of São Paulo as a financial center in the contemporary international division of labor.

In Chap. 2, we "go down one step" of the analytical scale, to understand financialization in the context of the Brazilian territory. Equally seeking to shed light on the role of the main information infrastructures that make up the national space, the chapter shows that, at the very beginning of the twentieth century, the financial structure of territory was more regionalized, founded on small and medium size banks of regional reach, and that were gradually being transformed into a more centralized financial structure, based on large commercial banks, and with its core in the Metropolitan Region of São Paulo. The chapter explains the recent development of the "banking topology" and the main aspects of the Brazilian market capital, as two main manifestations of the recent financial uses of the territory. Emphasis was given on commercial banks, including an examination of the new channels that also composes this "topology" for providing banking services, which made access to finance much easier for the population (even lower income population that lives far from the core areas of Brazilian economy).

Finally, Chap. 3 aims to identify the main financial uses of the territory on the local scale, highlighting important aspects of this low-intensive financialization process: how the low-income population has been inserted in the financial circuits recently, as well as to show how *credit cooperatives*, *community banks*, and *fintechs* are being developed in Brazil. We are considering these three types of financial agents as potentially disruptive forms of the Brazilian financial system, which is largely bank-centered. They can be viewed as alternative forms of provision of financial services and products, at least when compared to monopolist forms of this provision (and then, they can jeopardize the contemporary power structure of Brazilian financial system).

The structure of the argument divided into three chapters intends to initiate from broader historical processes of global breadth (Chap. 1) and then to perform more accurately national-scale (Chap. 2) and regional/local-scale analyzes (Chap. 3) of the Brazilian financial capitalism. Besides of being also "historical" and "theoretical," the last two chapters purposely have a more empirical and contemporary bias, aiming to better present the Brazilian financial geography to a larger public.

This book tries to understand these new realities also with a broader purpose: make the financialization of geographical space a mechanism of effective emancipation of nations and people, transforming the hegemonic uses of space in more public and democratic ones, focused on the necessities of most part of populations, especially the poor and subordinate populations of peripheral countries of the capitalist system.

References

Aalbers M (2015) Financial geography: introduction to the virtual issue. 40:300–305. https://doi.org/10.1111/tran.12081

Allix A (1950) Manual de Geografia General—Fisica, Humana y Economica. Ediciones Rialp, Madrid

Bennett RJ (1980) The geography of public finance. Welfare under Fiscal Federalism and Local Government Finance. Methuen, London

Braga JC (1985) Temporalidade da Riqueza. Uma contribuição à teoria da dinâmica capitalista. Instituto de Economia/UNICAMP, Campinas (Unpublished PhD Thesis)

Braga JC (1993) A financeirização da riqueza: a macroestrutura financeira e a nova dinâmica dos capitalismos centrais. Economia Sociedade 2(1):25–57

Chesnais F (1997) L'Émergence d'un régime d'accumulation mondiale à dominante financière. La Pensée no 309:61–84

Chisolm G (1889) Handbook of commercial geography. Longmans, Green and Co., New York

Christophers B (2015) The limits to financialization. Dialogues in human geography 5(2):183–200

Clarence Jones e Gordon Darkenwald (1941) Economic geography. Macmillan, New York

Engelen E (2008) The case for financialization. Competition Change 12(2):111–119

Epstein G (2005) Introduction: financialization and the world economy. In: Epstein G (ed) Financialization and the world economy. Edward Elgar, Chaltenham/Northampton, pp 3–16

Fix M (2007) São Paulo, cidade global. Fundamentos financeiros de uma miragem. Boitempo Editorial, São Paulo

Goldfarb Y (2015) Expansão da soja e financeirização da agricultura como expressões recentes do regime alimentar corporativo no Brasil e na Argentina: o exemplo da Cargill. Revista Nera 18 (28):32–67

Guttmann R, Plihon D (2008) O endividamento do consumidor no cerne do capitalismo conduzido pelas finanças. Economia Sociedade 17:575–610

Harvey D (1982) The limits to capital. Oxford: Blackwell Publishers

Hilferding R ([1910] 1985) O capital financeiro. Abril Cultural, São Paulo

Juglas JJ (1950) Économie Mondiale. Geographie Économique. Foucher, Paris

Kerr D (1965) Some aspects of the geography of finance in Canadá. Can Geogr 9(4):175–192

Keynes JM ([1936] 1964) The general theory of employment, interest and money. MacMillan and Co., London

Krippner G (2005) The financialization of the American economy. Socio-econ Rev 3:173–208
Labasse J (1955) Les Capitaux et la Région. Étude Géographique. Librairie Armand Colin/Cahiers de la FNSP, Paris
Labasse J (1974) L'Espace Financier. Analyse Géographique. Armand Colin, Paris
Lapavitsas C (2011) Theorizing financialization. Work Employ Soc 25(4):611–626
Lenin V ([1917] 1984) O Imperialismo, etapa superior do capitalismo. Editora Alfa-Ômega, São Paulo
Martin R (2002) Financialization of daily life. Temple University Press, Philadelphia
Paulani L (2009) A crise do regime de acumulação com dominância da valorização financeira e a situação do Brasil. Estudos Avançados 23(66):25–39
Paulani L (2017) Não há saída sem a reversão da financeirização. Estudos Avançados 31(89):29–35
Pike A, Pollard J (2010) Economic geographies of financialization. Econ Geogr 86(1):29–51
Sanfelici D (2013) Financeirização e a produção do espaço urbano no Brasil: uma contribuição ao debate. Eure 39(118):27–46
Santos M ([1975] 1979) The shared space: the two circuits of the urban economy in underdeveloped countries. Methuen, London
Santos M (1977) Society and space: social formation as theory and method. Antipode. 9(1):3–13
Santos M (1993) A Urbanização Brasileira. Hucitec, São Paulo
Santos M (1994) Técnica, Espaço, Tempo. Globalização e meio técnico-científico informacional. Hucitec, São Paulo
Santos M (1996) A Natureza do Espaço. Técnica e Tempo, Razão e Emoção. Hucitec, São Paulo (Translated to french in La Nature de L'Espace, L'Harmattan, Paris 1997)
Santos M (2000) Por uma outra globalização. Do Pensamento Único à Consciência Universal. Record, Rio de Janeiro (Published by Springer in English Toward an other globalization: from the single thought to universal conscience)
Santos M, Silveira ML (2001) Brasil. Sociedade e território no início do século XXI. Record, Rio de Janeiro
Schumpeter J (1997) A Teoria do Desenvolvimento Econômico. Nova Abril. Original work published in 1911, São Paulo
Sokol M (2013) Towards a 'newer' economic geography? Injecting finance and financialization into economic geographies. Camb J Reg Econ Soc 6:501–515

Contents

List of Tables

List of Maps

Chapter 1
Financialization as a Global Process

Abstract The first chapter shows how the structural conditions for the current process of financialization of the Brazilian territory developed. On the basis of these new conditions are the process of liberalization of the economy in practically all countries and the diffusion of technical infrastructures with global reach that have allowed for more efficient, broad, and rapid circulation of information and financial flows. Attention is drawn to the formation of the first financial centers and offshore markets—mainly in Europe but also offshore jurisdictions in more peripheral areas—as well as the relative position of the city of São Paulo as a financial center in the context of Latin America, and in the international division of labor as a whole.

Keywords International financial system · Global information networks · International financial centers · Neoliberalization · Latin America

1.1 From the Keynesianism of Bretton Woods to the Neoliberalism of Washington Consensus

As Braudel (1958) demonstrated in most of his work, the organization of geographical space and present time is mostly a result from the so-called *longue durée*. This is also true about the basis on which the world financial system and main international financial centers were created. This recent genesis remotes to the *Pax Britannica* period that occurred basically on the nineteenth century, more precisely from the year 1815, inaugurating what Arrighi (1994) named the "free trade imperialism."[1]

[1] Endowed with a powerful industry in all economic sectors that emerged in the wake of European bourgeois revolutions—mainly the industry of capital goods—England was also the main imperialist power in the world, with colonies spread in all continents, and that provided captive market to sell its products, in addition to large amount of resources from collecting taxes from these colonies (Arrighi 1994, 54). It should be reminded, however, that England was, since the end of the eighteenth century, the country with the most pronounced industrial and commercial development, becoming the great "exporter industry" of the world (Mantoux 1979). Regarding the so-called "Imperial Britain," even Mackinder ([1902] 1969, 343) also showed that "under a condition of universal free trade such a nation would almost inevitably increase its lead, and might ultimately reduce the whole world to economic subjection."

F. B. Contel, *The Financialization of the Brazilian Territory*, Economic Geography,
https://doi.org/10.1007/978-3-030-40293-8_1

This period was essential for the comprehension of the international financial system's current configuration, once at this time was formed the contemporary network of countries from British Commonwealth, which currently account for more than half of the amounts traded in international financial centers (including small British protectorates that operate as offshore financial jurisdictions, such as Isle de Man, Jersey Islands, Cayman Islands, etc.) (Palan 2010).

This geopolitical and economic power of England had its peak during the nineteenth century, and in respect to international finances, enabled them to dictate primitive forms of trade regulation and international exchange, creating an order based on gold convertibility and use of pound sterling for the international trade (the gold-pound standard) effective from 1870 until 1914 (Eichengreen 1995; Brunhoff 1996; Roberts 2000). The international monetary system operated as from fix exchange rates as standard.

In this context, the main discussions and decisions that influenced the economic arena—which was becoming more internationalized—depended mostly on what was decided in the British Parliament and in the internal committees of the Bank of England. The City of London concentrated already the main financial firms, stock exchange, and banks responsible for financing the national industry, the English international trade (Cassis 2006), including capital exportation operations for most of the peripheral countries, such as the case of financing railroads and commodities exportation in Latin America (Joslin 1963). Under these conditions, London was at that time the "natural center" of world's "high finance" (Arrighi 1994, 54).[2]

The 1929 crisis in USA was another central event for understanding the new order that was being formed, with repercussions for all history and geography of finance in the twentieth century. The crisis led the authorities of United States to radically change the regulation and operation of its financial system (mainly the banking system and capital market), implementing more rigorous measures to control the action of these agents. It was clearly noticed since then that the self-regulated market—as already identified a long time by Polanyi (1944)—produces catastrophic effects and generates major proportion economic crises, with widespread socialization of losses to the whole economic system. In case of US bank system, the main regulation measure was the implementation of Glass–Steagall Act on 1933, which resulted in the following changes: 1. segmentation of the operation of financial agents in the national credit market, separating the banks in different types (mainly investment and commercial banks); 2. implementation of minimum capital requirements for national and state banks; 3. prohibition of the operation of commercial banks outside their original area of creation, what allowed the maintenance of a more deconcentrated system in geographical terms and more decentralized in terms of its control (Fischer 1968).

[2]The great financial decisions made by the bankers from City were based on the necessity—or not—of maintaining the so-called "gold-standard" convertibility of the Pound Sterling (the main currency of international trade in all this period) (Roberts 2000). The so-called "bullionists," based mostly in the writings of David Ricardo, advocated about the convertibility necessity of English currency into gold, which obliged, therefore, the existence of a reserve of ingots in the Bank of England for each Pound issued (Leyshon and Thrift 1997, 64–66).

The feature of changes implemented by political and economic authorities of the main world economic power after the crisis of 1929/30—clearly non-liberal changes—helps us understand also the spirit that dominated the agreements of Bretton Woods. Largely conducted by the position of the United States and England, the agreement had as its basic ideological foundation the proposals of one of their main articulators, the British economist John Maynard Keynes (1883–1946). As shown by Leyshon and Thrift (1997, 74), "in common with the New Deal reformers, Keynes wished to restrict the mobility and freedom of financial capital to permit the construction of a more productivist industrial form of capitalist development." If, on one hand, the hegemonic countries were claiming for greater liberalism in the international arena (trade, investments, and regulations), paradoxically they advocated for more "interventionist" measures for the internal preservation of jobs and pursuit of full economic development of their territories. This development model became known as "Welfare State," and the international system may be comprehended since then as a welfare-nationalist-state model (Leyshon and Thrift 1997, 69), and had the following as main characteristics:

1. Use of State indebtedness capacity (mainly through the issuance/sale of public securities in the capital market) for investments in strategic sectors of economy;
2. These development/financial policies are mainly intended for job generation, especially in periods of greater economic crises (policies were clearly anti-cyclic, therefore);
3. Inflation was a natural, and not necessarily undesirable consequence of the stagnation/crisis's reversal period. As Cohen (1998, 43) affirms, Keynesian expansionist policies are efficient for unemployment reduction in sustainable form, even when implies in certain inflation pressure;
4. In practice, this policy model implied in an increase of economic nationalism and inversion of priorities when compared to policies of evidently more liberal character. As also stated by the Brazilian economist Belluzzo (1995, 12), in their initial plan, Bretton Woods Agreements aimed to "build an international economic environment destined to provide broad margin of manoeuvre for national policies of development, industrialization and social progress".[3]

From the point of view of the new institutionality of the international system created in this period, should be mentioned the creation of two important financial multilateral organizations, which would be fundamental for the organization of the world as we know today, and that are configured as the first set of systematic rules in history aiming "collective management" of international monetary market (Strange 1994, 55):

[3]Regarding the new rules preconized by Bretton Woods Agreements, it could be said that it was based on three main fundaments: 1. System of fixed exchange rates; 2. Currency convertibility (mainly from 1958); 3. Ostensibly national controls over employment, savings, and interest rates (Corbridge and Thrift 1994, 7).

1. Creation of the International Bank for Reconstruction and Development (IBRD) in December 1944—later known as World Bank—which would be a Bank destined for financing the reconstruction of the more affected countries immediately after the Second World War (mostly Japan and Germany). In medium term, other countries would also be object of IBRD concern, always by granting long-term loans (transferring "some funds from the developed world to the developing world") (Corbridge and Thrift 1994, 7). Soon the Bank would turn the attention to Third World countries, poorer in capital, and with large amount of natural resources and labor in their national economies and
2. The creation of International Monetary Fund (IMF) on December 1945, "set up to supervise this system and to assist countries in short-term balance of payments difficulties" (Corbridge and Thrift 1994, 7). IMF would be liable for promoting a more stable international monetary system, by financing countries with deficits in balance of payment, as to allow adjustments to be made without interfering in the national exchange rates (and that, therefore, would not allow the destabilization of international monetary system) (Strange 1994).

According to Leyshon and Thrift (1997, 72), Bretton Woods system carried two major contradictions that in the "medium term" of three decades would make the system collapse:

1. First, the fact that United States was at the same time the regulator country and the "lender of last resort" (based on the dollar transformation into the single effective international currency) and, at the same time, also an economic agent, an active competitor in the international financial system. The main concrete manifestation of this contradiction was the "capacity" of USA to solve severe problems of international payment with internal expansionist policies (dollar issuance) and
2. The regulation system preconized in Keynesian molds was based on the need of strong national monetary authorities, in a way that could implement economic policies in a more solid basis in their respective economic territories; but, at the same time, it had advocated an emphasis on liberalism in terms of international flows of capitals and external investment. This liberalism was in great part responsible for global diffusion of multinational companies, which undermined the possibility/capacity of national States from implementing economic policies in more autonomous and effective form (Santos 1975).

This way of thinking is also followed by Budd (1999, 115) which denominated this dualism of "embedded liberalism": on one hand, it was "tolerated" domestic state policies more "interventionists" (aiming full employment, increase of productivity, and general social well-being), and at the same time, it was necessary to maintain the commitment with free trade—and reinforcement of liberal policies—on the scale of international economy.

The Bretton Woods system was definitively disrupted in the year 1971, when the government of the U.S. President Richard Nixon broke the dollar convertibility into gold. How did this process take place?

On the 1970 decade, the US dollar was already consolidated as the main international currency, just as USA was the main economic, ideological, and military power in the world (Guimarães 1999). The diffusion of US multinationals and of the Fordist Mode of Regulation around the world, as well as use of dollar as main currency in international trade, cemented also US hegemony in the international system (Lipietz 1987). Because of its increased geopolitic power, and its greater control over UN and its agencies, USA starts to make an instrumental use of multilateral organizations, to solve their internal economic problems and to implement restrictions without precedents to "freedom of sovereign nations to organize the relations with other States and with their own citizens as they wanted" (Arrighi 1994, 70).

As to internal problems of USA, the main one was the huge deficit in commercial balance that USA started to present in the 1960', and that aggravated in the 1970'. With the industrial, commercial, and technological development of other economies (especially the rebuilded economies of Japan and Germany), USA started to demonstrate successive commercial deficits, increased by enormous military expenses made then, mainly with Korean (1950–1953) and Vietnam (1955–1975) wars, as Belluzzo (1995) shows.

According to Strange (1994), once United States guaranteed the gold–dollar convertibility, and maintained great amount of gold in their reserves, other countries were satisfied to have dollars as internal reserve in their central banks (instead of gold) (Strange 1994, 56). Everything was going well in the period in which the balance of payments of USA was positive and dollar shortage continued. If this system had been maintained, the internal gold reserves would soon become scarce, and the dollar credibility would be questioned, with aggravating fact that this race could take place abruptly, as in the case of bank going bankrupt (Strange, *op. cit.*, 57). As stated also by Hudson (1999, 142), to finance this deficit, more dollars were printed, what increased the mistrust of foreign dollar owners (and accelerated the race to exchange dollars for gold).

This unbalance is the basis for the collapse of the so-called "gold-dollar system," i.e., the parity of definition of dollar value with respect to gold, which was rigidly marked around US$ 35 per ounce of gold (Goldfinger 1986, 55). Because of the greater accelerated growth of the dollar mass in U.S. economy and abroad—when compared to gold reserves in the country Treasury—the parity, in the beginning of the 1970 decade, was only a "vision of the spirit" ("*vue de l'esprit*"), according to Goldfinger (1986). The industrialized nations already glimpsed this inability of the USA to maintain the parity, started a process—despite being slow—of exchanging dollars in their treasury for gold from US Federal Treasury. Aiming to break this possibility of depletion of gold reserves from FED, the Government of the President Nixon decided, unilaterally, to break the existing dollar-gold parity, no longer respecting the parity commitments that lasted from 1946 to 1971.

The break of gold-dollar standard in 1971, and consequently of the entire system set up in more regulated and "developmentalist" forms of Bretton Woods system, is one of the main bases for the current deregulation process of markets and financialization of economy, societies, and territories. Under the political point of view, the election of Margaret Thatcher in England (1979) and Ronald Reagan in USA

(1981) gave even more power to the diffusion of neoliberal ideology in all discussion forums and multilateral institutions, consolidating what Chesnais (1997) called "conservative revolution" in the international system.

In addition to greater mobility of financial flows provided by new transportation and information technologies (as will be seen in the next item), at the end of the 1970 decade also began a vigorous process of deregulation of the world economic system, which accelerated the mobility of flows and investments, increased global economy instability, and allowed formation of new financial territorialities (such as offshore financial jurisdictions). This deregulation implied in flexibilization/liberalization of national exchange policies as well, which allows fluctuation of national currency values, what caused an even greater market instability (both in foreign exchange market and in new financial products associated to it). This instability and its use by a great number of institutional investors to profit with speculation assets on global level were essential factors for the creation of this tenser and more uncontrolled world, which Strange ([1987] 1997) named *Casino Capitalism*.[4]

In the words of Belluzzo (1995, 18), the national States, especially from peripheral countries, observe powerless:

> the deployment of localization strategies and internal division of labor of major company and is each time more subject to the tensions generated by the financial markets, which are subject to the whims of monetary, fiscal and foreign exchange policies. More than for its global character, the new finance and its logic became decisive due to its capacity to impose vetoes to (national) macroeconomic policies.

Beyond this greater mobility of finances, other fundamental consequence of this substitution of a Keynesian economic model for another more neoliberal was sensed by the peripheral counties of world system, mainly in Latin America. This new characteristic of world financial and monetary system was on the basis for a deep process of indebtedness of Latin American nations—mostly the three greatest economies in the region, Brazil, Mexico, and Argentina—with disastrous consequences that until now prevent these countries from finding autonomous and efficient mechanisms for their economic and social development (Sunkel 1986; Toussaint 2002).

Since the middle of 1960 decade, the increase of oil exportation and the surplus generated by these flows to great producer countries (especially from Middle East) made available a gigantic amount of financial resources, arising from successive and generous trade surplus on their commercial balances. The countries from OPEC became poles of financial wealth, after successive price increases of these commodities along the entire 1960 decade (and mainly after the expressive rise from 1973). It was necessary to "recycle" these substantial resources, once the economies where

4

> As in a casino, the world of high finance today offers the players a choice of games. Instead of roulette, blackjack, or poker, there is dealing to be done—the foreign exchange market and all its variations; or in bonds, government securities, or shares. In all these markets, you may place bets on the future by dealing forward and by buying or selling options and all sorts of other recondite financial inventions (Strange [1986] 1997, 1).

they were generated did not allow to be reinvested (due to the low complexity of these economies, and because of traditional political structures existing there).

The main destination of these surpluses was exactly the "offshore" market that was being formed in Europe, the so-called Euromarket, which offered more interesting interests/remuneration, lower fees for the services and guarantees of anonymity that the "traditional" onshore markets did not offer. The Euromarket—which is one of the basis of the financialization process of world economy—may be defined as a financial market created from the 1950 decade, in countries that started to perform financial operations denominated in non-resident currency, especially dollar; it consists, therefore, in a market where the transactions were primordially made with "Eurodollars," i.e., the dollars that were outside the US territory and monetary authorities. The prefix "Euro," in fact, means simply "offshore," out of the control of US authorities (Budd 1999, 125), and therefore the term also causes certain confusion (once remits only to the European continent, when, in fact, this type of market may take place in any country—or jurisdiction—that allows financial operations similar in their national space) (Clark et al. 2015).[5]

These large resources that were being deposited at the main financial European markets, in turn, needed to be reinvested in other places. "They found them in many developing countries hit by high oil prices and hungry for economic development" (Strange 1994, 98). These resources, different from what occurred in the World Bank action right after World War II, had few conditions or clauses/performance requirements. Added to the growing world inflation from the middle of the 1970 decade, the real cost of these loans was very attractive. "It was all too easy," reminds Strange (1994). It is also important to remember that in this period most Latin American countries (including Brazil, Argentina, Chile, Uruguay, and Paraguay) were under dictatorial military political regimes, which also prevented any wider public/democratic control over this foreign indebtedness process (Sunkel 1986). In this scenario, and in the incapacity of the multinationals in expansion to absorb this excess of liquidity in the European banks, these same banks target these surpluses for loans to national peripheral States, "in the form of huge loans to less developed countries in the Third World, particularly Latin America" (Martin 1994, 258).

The problem only appeared after the crises of 1973. The main issue arised from the fact that these interest rates charged from most of the loans were variable. "Loans were made at variables rates, at so much above the interbank rate in London, LIBOR. If LIBOR went up, so did the burden of servicing (i.e., paying interest) on the loans" (Strange 1994, 58). After years supplying the Latin American economies, with the oil crises (with great increase in its cost in 1973), the flow of resources inverts, therefore.

---

[5] As the author states, "It was in this period that international banks started opening deposit accounts denominated in a currency different than that of the country in which the bank was located. Branches of U.S. banks based in London led the trend, running U.S. dollar accounts out of London, not subject to U.S. regulations. Accepting deposits in foreign currencies was followed by lending, and issuance of bonds and other financial instruments. As such the term offshore originally referred to currencies and was related to the development of Eurocurrency markets. Offshore financial markets were Eurocurrency markets, with leading centers in London, Luxembourg, and later Singapore" (Clark et al. 2015, 240).

As demonstrated by Charles Goldfinger, if until the 1980 decade (especially until 1982, before the Latin American debt crisis) the flows were directed to the periphery, as from which, there is a "return to sources," the banks from rich countries become suction pumps of resources in periphery (Goldfinger 1986, 97).

This is, schematically, the general scenario of international economy that was developed throughout the entire second half of the twentieth century. The more social, democratic, and developmentalist spirit of the monetary and financial system initially created in 1945 was gradually losing power, and the last catalyzer of the substitution process of the original model of Bretton Woods was the acceleration of the global diffusion of neoliberal ideology (Anderson 1991). As shown by Peck (2001), one of the main forms of institutionalization of this neoliberal ideology was through the creation of the Washington Consensus, a document published in 1989 by the *Institute for International Economics* (under the primary responsibility of John Williamson), with the purpose of "naturalizing" certain types of public policies, as if they were a necessary and inescapable consequence of globalization, and that therefore should have been fully adopted, in all countries of the world (Peck 2001, 448/449). The main measures are Fiscal discipline, Tax reform, Financial liberalization ("interest rates and capital flows should be market determined"), Liberalization of Exchange rates, Trade liberalization ("quantitative restrictions on imports should be removed"), Foreign direct investment ("barriers to the entry of foreign firms should be abolished"), Privatization ("state enterprises should be returned to private ownership"), and general Deregulation ("governments should abolish regulations restricting competition") (Peck, *op. cit*).

The neoliberal recipe was also strongly disclosed once it becomes part of the conditions imposed by the two main multinational financial institutions, IMF and World Bank (Toussaint 2002; Freitas 2011). For these two institutions, the opening of market and denationalization/privatization—mainly of national banking systems—would strengthen the systems to resist economic crises, as well as would introduce greater competition and improvement of domestic banking systems (Freitas and Prates 2000, 63).

This liberalization, as already mentioned, is one of the main causes of the contemporaneous process of financialization of the global space. Jointly with the diffusion of information technology in global scale, the liberalization process has enabled financial institutions to create new products and services (or financial innovations), increasingly difficult to be regulated by national jurisdictions/States (O'Brien 1992; Guttmann 1996). For Arroyo (2006, 185), this process enlarged the "porosity" of the economic frontiers of Latin American countries to global financial capital, also increasing their external vulnerability.

Therefore, the mobility and reproduction conditions of international financial institutions turn themselves practically ubiquitous (O'Brien 1992). Money becomes "hypermobile" (Warf 1999), and together with the development of a complex market for new financial products and services, inaugurated what Goldfinger (1986, 82) denominated *geofinance*, this new space-time which is the material and political basis of the financialization process in its contemporary feature. In fact, in the long run, since the break of the gold standard in the begging of the twentieth century, money itself start to become more *dematerialized*, and more fluid and instable (Brunhoff 1967, 1996). This dematerialization increased when President Nixon ended dollar convertibility to gold in 1971, creating what Kurtzman (1994) called the "megabyte money"—or cybercash, according to Guttmann (1996)—a new kind of money that is in the basis of the nowadays financialization process. This new form of money, in turn, could not exist if there had not been constant and ever wider changes in the geographic infrastructures that allowed the transactions—and the very existence—of this new money and financial instruments, as we will see in the following item.

1.2 The Technical Basis of Financialization: The Global Reach of Information Networks

All this political–economic process described in the previous item was supported by a profound transformation of some technical infrastructures with international reach and also by the creation of "large technical systems" (Hughes 1983; Gras 1993), within and between national territories. Gradually, since the end of the nineteenth century, was being diffused, selectively, but constantly, a new geographic *milieu*, as observed by Santos (1996). This new geographic *milieu* that was more intensive in technical and scientific innovations, for starts, and that gradually incorporates also information techniques, transforming it into a "technical-scientific-informational *milieu*" (Santos 1996).

With the development of information techniques that occurred mainly as from World War II, new economic sectors and firms were created, as well as were added an entire new family of information technologies to industrial plants, commercial buildings, railways, roadways, ports, etc. Gradually data storage and processing instruments become more powerful and efficient, and the link between these machines and transmission networks creates what French sociology named "*la télématique*" (Nora and Minc 1978; Mathelot 1982). For Castells (1996), microelectronics, computers, and new forms of telecommunication founded a new economic paradigm and are the basis of the "network society" in which we live to this day. What were, however, the first global information infrastructures that emerged in the context of the creation of an international financial system?

As regards the transmission of information, initially permitted by the telegraphy and broadcasting techniques, the information techniques became more sophisticated, more efficient, and increased their geographic reach to all earth surface (Bakis 1984, 1987; Sola Pool 1993). The telephone transmission networks, the optic fiber networks, the telecommunications systems by satellites—and more recently the proper Internet—made the information techniques one of the key variables of the current historical period, creating a completely new world "technical unicity" ("*unicidade técnica*"), as observed Santos (1994a, 1996).

These technical transformations were essential for the creation of great internal markets within the rich countries and for the expansion of their influence in the peripheral countries of the world system that also become integrated into the system assembled by even broader transportation and communication networks (Mattelart 1992). In addition to the internal "spatial fix" that these techniques allowed, the capitalist superpowers expand to the peripheries their needs and vicissitudes (Harvey 1982).

In the case of financial market—that also started a more robust internationalization process—, the telegraph technologies in a first moment, telephone and radio in a second, and of fiber optics and satellites in a third phase were central for its expansion and internationalization. As also demonstrated by Warf (1989, 261), "the financial sector is highly information-intensive, and it would be impossible for banks, securities firms, and insurance companies to provide the services they do without extensive telecommunication systems."

The diffusion of these material information networks is both cause and consequence of new international division of labor that are developed and consolidate the political and economic hierarchy of countries as from large metropolis from the center of capitalist system. The diffusion of a global submarine cable network allowed the logic of financial corporations to spread worldwide, mainly from the great nodes of the world urban network (Scott and Zachariadis 2012, 3).

As shown by Ash (2014, p. 20), the periodization of these basic infrastructures (or *réseau-support*) may be defined as follows:

1. Era of telegraph networks (from 1850 to 1950);
2. Era of telephone networks (from 1950 to 1986);
3. Era of optical—or digital—networks (from 1986 to current date).

Differences in the technical constitution of these networks and equipments, as well as in the availability of them at places, are definitive factors to change the velocity of transactions, reduce costs, and provide greater efficacy for the action of economic agents (mainly financial agents). These transformations allowed by the information infrastructures, on its turn, lead to an unequal development of regions and nations. In other words, the unequal distribution of the technical-scientific-information milieu is one of the conductor factors—or inhibitors—of economic development of nations and dynamism of regions (Santos 1994a).

The telegraph era, as mentioned (Ash 2014, 21), starts in 1850 all the way to 1950. It was fundamental once in fact revolutionized the long- and medium-distance communication, previously based on methods that depended on rudimentary systems, attached to natural resources and/or animal traction (use of carrier pigeons, horses, carriages, etc.). The creation and spatial diffusion of telegraphs enables for the first time an effective "uncoupling" between the information transmission and its physical base, what allows a wider spatial or social diffusion of information (Mattelart 1992). Until then, the information to be transmitted should be registered on material basis—especially papers—and transported inside mail bags or boxes, more or less large, according to the amount of information. The capacity, speed, and geographic amplitude of these systems were expressively lower than what became the national and international networks of telegraphs. The greater difficulty in transmission of information, in turn, made the context of action of companies, public organizations, and individuals much more localized or regionalized; as would say the geographer Werlen (1993), the context of actions prior to the modern techniques of transportation and communication was more "anchored" (Verankerten).

Leyshon (1995, 25) shows that the telegraphs quickly took over essential economic, but also geopolitical functions. Under the economic point of view, it should be remembered that the main users were stockbrokers and businessmen (mainly traders). In respect to its political importance, it was essential for the communication between diplomats and the national States, especially those with great ultramarine dependencies. A third function is added to the two previous ones: work instrument for journalists. In this sense, it is not to wonder that the main world power at the time—England—was the first country to use systematically its telegraph network for the geopolitical control of its vast imperium (Malecki and We 2009, 361), and also developed the first news agency of international character, Reuters (Leyshon 1995, 27).

This technology very early influenced also financial markets. In the mid-1850, English investors had US debt securities, which in turn were sold on both New York and London stock exchanges. The information transmission time about the quotation of these papers was the time that ships took to cross the Atlantic sea, around 3 weeks. The investors that sold and bought papers from USA needed to estimate the prices that should be practiced in that market, and even when the information about them arrived, it would be—again—3 weeks delayed. With the installation of the first transatlantic cable in 1866, this delay falls to 1 day. Increases precision about the price estimates, and the operators start to work with a much greater accuracy level (Garbade and Silber 1978, 820).

This first transatlantic international telegraph submarine cable, installed between New York and London in 1866, allowed a greater integration between securities' markets existing in both cities. Its use on the stock exchanges was quick, and almost "immediately" the use allowed to reduce the stock price differential between New York and geographically disperse regional stock markets (Batiz-Lalo and Woods 2002, 3).[6] On the year 1870 started to operate the cables that would connect London to Bombay and Madras. On 1871, this network would reach Penang (Malaysia), Singapore, and Hong Kong (through Saigon). At the end of 1872, the telegraphic network of the British Empire would reach Darwin and Adelaide, in Australia (Ash 2014, 24).

[6]It was uncommon at that time to use communication by telegraph for speculative purposes—or for agreed price determination between actors. Added to this the proper characteristic of US economy—especially its monetary system—still not totally integrated to the beginning of twentieth century. Those quotation distinctions in the different cities of the country generated an exchange market inside the US economy. The economic agents had to consider the currency cost in the markets that acted—outside their own—when at the moment of performing a commercial transaction in another market (Garbade and Silber 1978, 823). New York and New Orleans were the main exchange markets in the country. The relation between the two cities was extremely dynamic, especially due to the production and exportation circuit of cotton (that exited from the south and was exported to England, especially through the port of New York). (Garbade and Silber, *op. cit.*, 824) The telegraph technology promoted an effective integration of the exchange market in USA. (*op. cit.*, 825).

The first transpacific telegraphic cable, on its turn, begun to be constructed in 1896, and was concluded in 1902; the cable started in Australia, and through New Zealand reached Norfolk Island, Fiji Islands, Fanning Island to finally arrive in Vancouver, Canada (Ash 2014, 27). With 3458 nautical miles, it was the greatest cable installed in the world at that time. In this same period, the US company Commercial Pacific Cable Company installs also a transpacific cable that would connect the city of São Francisco, through Honolulu, to Guam and Manila (in Philippines). These two cables allowed for a telegram to be sent from London to Sydney in only one hour, a true revolution at that time (Ash 2014, 27).

Since 1866, however, with the more accelerated diffusion of several submarine cables, the main cities, in all continents, became connected by this new support network. The information transmission speed grew exponentially.[7] The new information technologies affect also the operation of banking system, mainly in case of USA and Great Britain. As shown by Batiz-Lalo and Woods (2002), the introduction of these innovations associated to telegraph inaugurates what may be called as "Early adoption period," that took place from 1864 to 1945, and would dictate the rhythm and form of diffusion of banking activity in core countries of capitalist system.

For Garbade and Silber (1978), this new technical basis was one of the main factors to cause a "disruptive" transformation in the US and world financial market. This diffusion of this new technical system within US territory and the installation of a line between USA and England generated three main "gains" for the financial operators:

1. "an equalization of securities prices and hence an equalization of the marginal rates of substitution between different securities. Thus, there are gains to the direct consumers of the financial market—portfolio managers" (Garbade and Silber 1978, 822);
2. Gains for real economy (consume and investment);
3. By uniting separate markets, "financial markets provide liquidity services as well as influencing portfolio allocation and real sector decision-making. [...] Innovation of faster communications reduces the costs of searching the best price in alternative markets" (*op. cit.*, 822) (or "reduced search costs").

It is important to also mention that this more broad diffusion of submarine cables around the world and the increase of the number of companies that also starts to install their own cables (in addition to the English, also French and North American companies), have caused the transmission costs to become lower. For a transmission of at least two existing transatlantic cables, the rates that were around 40 cents per word reached 12 cents, and stabilized at a price around 25 cents per word, in the year 1886 (Ash 2014, 26).

At the end of the nineteenth century, two other technical innovations started to compete with the telegraph cables: the radio diffusion systems in short waves and

[7] As a comparative example, the news of the murder of Lincoln in 1865 took 15 days to reach London; in 1891, the earthquake in Nobi (Japan) was known one hour after the event (Malecki and We 2009, 361).

telex. Used initially for communication between the stations in Great Britain and the boats from the military freight of the country, in the year 1919 the telegraph technology by radio becomes commercially feasible and competes directly with the cable transmission.[8]

In 1929—also due to the crash of New York Stock Exchange—the radio technologies have spread with great impetuous, reducing the costs for information transmission to almost 1/6 of the cost for cable transmission; in 1937, as shown by Malecki and We (2009, 361), the number of international wireless circuits is almost equivalent to the global network of submarine cables.

The second great innovation also in nineteenth century that improves information transmission—including financial information—was the Telex, a system that would allow to insert information in keyboard with letters and numbers (typewriter keyboard) that projected information on screen (teleprinter exchange). This technology was developed by German companies (especially *Siemens and Halske*) in the context of the Second World War and was promptly adopted by England, France, USA, and Canada (Scott and Zachariadis 2012, 3).

It should be observed that practically all innovations in this period (from telegraph to telex and teleprompter) occurred fundamentally because of the preexistence of network infrastructures (réseau-support) which the new system also used. As the authors remind, the teleprompter technology "initially based on the use of the existing telephone and telegraph networks and allowed speech and teleprinter signals on the same connection" (Scott and Zachariadis 2012, 3). The large corporations become the main demanders of the Telex technology, with emphasis on financial firms (major banks, stock exchange, and other operators). With the use of these systems, banks reduced their operation costs (avoiding also use of paper for transaction effectiveness) and increased their operational efficacy (Scott and Zachariadis 2012, 3).

This standard based on the mix of telegraph cables and radio broadcasting will be maintained until mid-1950, when the first telephone cables were developed, which quickly made obsolete telegraph cables. The first transatlantic telephone cable installed started operation on 1956, the Transatlantic No. 1—TAT-1 (Malecki and We 2009, 361). The TAT-1 consisted of one coaxial cable isolated with polyethylene and used vacuum as repeaters. It connected the city of Oban (Scotland) to Clarenville (in the providence of Labrador, Canada). It was built by the financial investment from three main companies: the North American AT&T, the Canadian Overseas Telecommunications Corporation, and the British UK General Post Office. As shown by Eichengreen et al. (2016, 11), "When inaugurated on September 25, 1956, it had 36 separate channels, enabling it to carry 35 simultaneous telephone calls along with 22 telegraph lines on the 36th channel."[9]

[8]As shown by Ash (2014, 27/28), submarine telegraphy continued into the 1960 and, although some improvements continued to be made in cable design and transmission techniques, the rise of radio technology meant that the industry ground to a halt and then went into decline awaiting a paradigm shift in transmission technology.

[9]One of the main technical innovations was the introduction of coaxial cables in long-distance data transmission, technology that offered more load capacity, more efficiency/reliability, and that operated transistors—instead of vacuum tubes as repeaters (Eichengreen et al. 2016, p. 11/12).

Because of the rupture brought by this new technique, Ash (2014) argues that a new era in the international telecommunications is launched, the era of telephonic networks (1950–1986). After increased technical efficacy and after becoming relatively cheaper to be produced and installed, the American and British companies launched the diffusion of longer submarine cables (Transocean cables). After the technical and commercial success of TAT-1 (with capacity for voice signal transmission in 36 × 4 Kz), the second great telephone cable installed was CANTAT-1 (Canadian Transatlantic Telephone Cable), installed in 1961, already with a technology of repeaters more appropriate than TAT-1. The CANTAT-1 connected the city of Corner Brook (Canada) to Oban, in Scotland. (Ash 2014, 31). The first cable in the Pacific Ocean to be installed was COMPAC, installed in the year 1963. The cables had capacity for transmission of 80 × 3 Kz voice channels and used 322 repeaters in the entire extension. It connected Australia, New Zealand, Fiji Island, Hawaii, and finally Canada. They were also built by SCL and by the Standard Telephones and Cables ltd (STC) (both British companies).

Since the installation of COMPAC, the cable transmission capacity grew significantly. Improvements in the proper cables, as well as in the technology of repeaters, allowed for the systems that usually work ranging from 5 to 12 MHz to soon reach 30, 36, and finally 45 MHz (Ash 2014, 31).

The Cable PENCAN-3, installed in 1971, was also a type of hallmark in technological terms for the time: it contained a super unit of 5680 × 3 kHz voice channels, representing over 23 times the capacity of the cables installed in the previous years. The greater densities of cables between Atlantic and Pacific oceans reflect the telecommunication necessities between the main economies in the world (and their global cities), making these two macro-areas the two greatest markets for transmission of intercontinental/global information (Warf 2006). According to Ash (2014, 32), the last Transocean cable installed was TAT-7, in 1983, that provided 4246 × 3 kHz voice channels, using 67 repeaters in sequence, distanced 9.46 km, one from the others.

In respect to operation of the banks—especially in USA—these new information transmission techniqùes—telegraph, telex, and telephone—allowed the bank headquarters to communicate with more efficacy with branch offices/agencies, enabling centralization of organization control (which results in much larger economies of scale), as well as the possibility of better "distribution of funds" jointly with the geographic markets/areas in which the banks operate (Bátiz-Lazo and Wood 2002, 3). Among the main changes occurred are highlighted as follows:

1. Accounting control of network of branches and subsidiaries as a whole, since the headquarters started to receive more efficiently the aggregate results of its branches.
2. In general, the headquarters/central branches also managed:
    a. check clearing,
    b. the relations with Central Bank,
    c. execution of treasury operations.

3. The headquarters also perform more stringent supervision and financial control of branches, often using "draconian inspection methods."

The adoption of these techniques permitted—and was also catalyzed by—the growth of the number of bank branches, and therefore of their branch subsidiaries. The improved communication enabled to identify agencies and managers that did not reach the performance desired by the corporation (Bátiz-Lazo and Wood 2002, 3).

In mid-1980 is inaugurated what would be the "last"—or most recent—phase of international telecommunications, which we currently are: the fiber-optic era (Ash 2014). As Barney Warf (2017a, 113) shows, "fiber optics are long, thin, flexible, highly transparent rods of quartz glass (or less commonly, plastic) about the thickness of a human hair that can transmit light signals through a process of internal reflection, which retains light in the core and transforms the cable into a waveguide."

The cables may efficiently transmit voice, video, and data, in general, at the speed of light (300.000 km/s), making it, therefore, the fastest—and with greater capacity—for information transmission in contemporary world. The most modern optic fiber cables consist of around 1000 micro-size cables and are ideal to transmit "point-to-point" information (Warf 2017a, 113/114). The already "outdated" copper cables are gradually being replaced by optic fiber cables, mainly by telephone companies. As demonstrated by Warf (2017a, 115), this new material basis was an important catalyzer for the development of a series of other economic activities, mainly those related to trade sector, services, and finances.

At this time, the development of orbital satellite technologies starts to threaten the primacy of submarine telephone cables as main long-distance telecommunication means. The advantages that the cables still presented were: 1. they had lower latency in transmission; 2. produced less echo; and 3. are safer (Ash 2014, 33). Even with these conditions, as stated by Ash (2014, 32),

> during the 1970s and early 1980s satellites gradually became the predominant mode of transmission for international telephony. In 1986 the telephone Era came to an end with the laying of the last Stc 14 mHz system between India and the United Arab Emirates and, in the same year, the deployment of the first fiber optic systems.

It was in this new geographical *milieu*, more intensive in information technologies, that were created new service networks exclusively dedicated for use by financial corporations. The most important one, with a global reach, is the Society for Worldwide Interbank Financial Telecommunication (SWIFT). Founded in the city of Brussels in 1973, SWIFT assisted at that time the conduct of financial operations between 15 countries (basically those comprising the European Economic Community) and had 239 banks as clients so far. Charles Goldfinger defined it in the mid-1980s as a "pure interbank network, which ensures a telematic connection between 1,200 banks in forty countries" (Goldfinger 1986, 291). Currently, according to the institution's annual magazine, there are approximately 11,000 financial agents using SWIFT services in 200 territories (SWIFT 2018). As the institution's website shows, their "technologies have previously disrupted manual processes and unlocked efficiencies in the financial system, dramatically reducing frictions, costs and operational risks."

As Dörry et al. (2018) state, because of its importance to the international financial system, SWIFT has become a "critical service provider" for interbank messages and transactions of various sizes, as well as operating in other financial markets.[10]

The two most important national financial service networks in the Anglo-Saxon world are Bankers' Automated Clearing Services (BACS) that serves British banks and Clearing House Interbank Payments System (CHIPS) which serves mainly large US banks (Scott and Zachariadis 2012, 3). BACS is a non-profitable private organization, created in 1968, which operates as from its associated members and is responsible for "clearing and settlement of automated payments in the UK." BACS is owned by 15 of the main British—and also European—banks. This system is used by financial operators to do "the vast majority of credit transfers, standing orders and direct debts" (Kirkman 1987, 221). CHIPS is a privately held organization headquartered in New York that "serves the needs of large financial institutions operating in the United States and dealing in US dollars" and started their activities in the year 1970 (Kurtzman 1994). At the time it provided service for 9 US financial institutions, and in 2010 this number was 48 (mainly banks). In more practical terms it may be defined as "bank-to-bank electronic transfers" company (Scott and Zachariadis 2012, 15).

It is important to note that at that time the banks from core countries start a more substantial internationalization, once the information technologies made feasible this greater efficiency for exchanging data and messages between units around the world. The Transnational Banks (TNBs) emerged, which expanded at the same pace of the internationalization of the financial system. These banks also followed the diffusion of all types of non-financial transnational corporations, which expanded in the same period. As shown by the Scott and Zachariadis (2012, 3) "as large multinational banks turned their attention to establishing global operations and developing international business, their requirements for reliable computerized communication systems increased."

Under the technical point of view, the greatest innovation of the period was the use of Electronic Data Interchange (EDI) Systems, the purpose of which was to integrate financial and non-financial companies, allowing banks to increase the types of products and services offered worldwide (Scott and Zachariadis 2012, 3). These private networks were seen as strategic by the banks, given the fact that the public networks were considered insufficient (mainly due to their respective transmission capacities) and could become an embarrassment for the increase of financial transactions volume that grew widely. (Scott and Zachariadis, *op. cit.*, 3). Among the banks that were in the "front line" of these investments may be cited Citibank N.A., Bank of America and Chase Manhattan of USA, as well as Barclays, Lloyds, and Midland Bank in United Kingdom. (*idem*, 3). These large banks became pioneer in the development

[10]For the authors, "More than a mere passive facilitator of economic activity and a neutral provider of a level playing field, today SWIFT is a global monopoly highly sensitive to geo-political upheavals in the world. At the same time, it is increasingly influential and seemingly prone to act to the benefit of the world's most powerful financial and political players" (Dörry et al. 2018, 12).

of "proprietary private networks," renting data transmission capacities of other networks and satellites, mostly of companies—public and private—connected to the so-called PTT circuits (Postal, Telephone, and Telegraph) (Scott and Zachariadis 2012, 4).

As the demand for these infrastructures for information transmission increased, investments were also made in the construction of fiber-optic networks. Fiber optics spreads stronger and is a technology that allows much faster and cheaper transmission of information. As stated by Leyshon, with this new material basis, "money can now travel as electrons along telephone cables, at the speed of the light as photons long fiber-optic cables, and as pulses of energy transmitted via microwave and satellite communication systems" (Leyshon 1995, 35).

The beginning of the fiber-optic era took place as from the installation of the first submarine cables of this type in 1986, in the following regions:

1. Cable between Honshu and Hokkaido Islands in Japan (a 300 km domestic line);
2. In the same year, Alcatel installs their first system, connecting France to Corsica;
3. AT&T SSI installs an experimental system called Optican-1 at Canary Islands;
4. Finally, the British STC installed the first fiber optics international system, connecting United Kingdom to Belgium, UK–Belgium no. 5 (Ash 2014, 34).

The first transatlantic cable fully built and operated with the new fiber optic technology was TAT-8 in 1988. Despite the unimagined increase in the capacity of cable system provided by optic fibers, security problems with external risks also increased. These problems were due to the greater amount of cables installed on the floor of the oceans, and the increase of ships and fishing activity (reminding that anchors and fishing nets at high sea are the greatest "villain" for the integrity of the submarine cables).[11]

The modernization of these large intercontinental transmission lines and the increase of information transmission capacity that fiber optics brought have made submarine cables, once again, more interesting than the artificial satellites. As shown by Ash (2014), the first-generation optic repeaters operated with a transmission wavelength of 1310 nautical miles and digital transmission rate of 280 mbit/s. In 1990, the technology progressed and the transmission wavelength that was the basis of TAT-9 increased to 1550 nautical miles, with a digital transmission rate of 565 mbit/s, allowing nothing less than the transmission possibility of voice channel 80.000 $\times$ 64 kbit/s. This already overcame the capacity available by satellite and once again "submarine cables became the dominant international telecommunications medium" (Ash 2014, 35).

Probably the main financial activity that is affected by the speed increase provided by the digital technologies is the exchange market. The foreign exchange market is the most sensitive of financial activities with respect to the transaction time issue. The study of Eichengreen et al. (2016) focused its analysis in the three main world

[11]The solution found to reduce aggression risks was the use of "plows" at the moment of installing the cables, as to bury them at coast areas; as sown by Ash (2014, 35) "it is now standard practice to bury cable in water depths up to 1000 m, and sometimes beyond, to protect them against fishing activity."

financial markets: New York, London, and Tokyo, once the main servers are installed in these marketplaces, and also the two main foreign exchange brokerage and trading companies: the Electronic Broking Services (EBS) and Thomson Reuters. The connections of an economic agent to United Kingdom (for London), to USA (for New York), or to Japan (for Tokyo) are, therefore, made by fiber-optic submarine cable (which reduces the latency time and increases the bandwidth). Regarding foreign exchange markets, therefore, these technological innovations may have two main effects in the localization of activities:

1. They can increase the power of main global financial centers, once they may increase the preference of agents in performing businesses in these centers, because the speed there would be greater, and therefore the profitability of business; and
2. At the same time, it may be argued that the reduction in transaction costs provided would allow the retention in the issuer market of a greater part of the transactions (Eichengreen et al. 2016, 3).

Another essential aspect of this new era of information transmission is that the costs associated to this phenomenon decrease at even greater levels. As shown by Eichengreen et al. (2016, 5), "the cost of everything related to information and communication is dropping like a stone due to developments in computation, big data, fiber optics and satellite technology." Currently, as stated by Warf (2017b, 398/399):

> This infrastructure emerged as the backbone of the global financial system and played a powerful role in the growth of the internet. Using fiber optics, which greatly exceeded the capacity of competing systems such as satellites, financial institutions designed electronic funds transfer systems (EFTS) to shift funds effortlessly across borders, take advantage of differences in interest rates, speculate on exchange rate fluctuations, and avoid political upheaval.

The dynamism and "spatial productivity" (Santos 1994a) of cities and financial operators are therefore dependent on the availability of modern infrastructures of information transmission, and those more modern systems with greater capacity give important economic advantages to the operators who settle next to them. The distribution and qualities of the infrastructures, therefore, are elements that condition the contemporary financial dynamics for places and regions. In a more political perspective, Warf (1999) points that the diffusion of these global information networks was, directly and indirectly, one of the main causes of the increase in the power of large global companies and the consequent collapse of Keynesianism and the worldwide diffusion of neoliberal ideology.

1.3 The New Territoriality of Financialization: Offshore Financial Centers

Along with the creation of the Euromarket and the diffusion of a new global support network by satellite systems and submarine cables, another essential phenomenon of the financialization that took place from the 1960s and 1970s was the development of Offshore Financial Centers (OFC) or "offshore jurisdictions" (Wójcik 2012a, Clark et al. 2015). These new geo-economic forms allowed even greater freedom for cash flows and for the global performance of all types of firms, especially multinationals. These centers were also essential for large companies from all economic sectors to develop aggressive strategies of tax evasion and avoidance, as well as reorganization of their internal financial operations.

Some of the practices that today characterize offshore financial centers are old. Several countries, or jurisdictions within countries, have specialized throughout history in commercial and financial activities by enacting laws that favor these activities. Free ports, free trade zones, unguarded roads and routes, and cities with a large presence of money changers and merchants have for a long time enjoyed the fiscal "lack of regulation" that characterizes the so-called "tax havens." At present, it is possible to identify a complex division of labor between them, including a central European area (consisting mainly of Ireland, the Netherlands, and Luxembourg) and a set of other more peripheral tax havens (Crofts and Zigler 2018).

To differentiate tax havens from offshore financial centers, the former are locations that developed prior to the period of globalization and the massive introduction of information technology in financial affairs. As Vernay (1970) points out in his book *Los paraísos fiscales*, these "spinning platforms of money" have been created throughout history in places with a privileged geographic location relative to international trade flows, usually port cities or major junctions inside territories. In these places, local authorities guaranteed very liberal regulation systems for the companies that settled there, favoring mainly financial and commercial activities, whether legal or illegal. These are the most general reasons for the development of the financial markets of European cities and countries such as Monaco, Liechtenstein, Andorra, Luxembourg, and Switzerland, but also true "*enclaves*" of free trade and regulation in peripheral countries, such as Tangier (Morocco), Monrovia (Liberia), Macau, Beirut, Panama City, as well as islands like Bahamas, Guernsey, and Jersey (Vernay 1970).

The tax havens are not easily distinguishable from what currently characterizes offshore financial centers simply because the main features of tax havens are repeated in today's offshore financial centers (Roberts 1994, 99). Most offshore financial centers in the 1970s were formed from what used to be tax havens. As shown by Cobb (1998), we can distinguish three general causes that explain the emergence of offshore financial centers from the 1960s and 1970s:

1. Demise of the Bretton Woods system of fixed exchange rates;
2. Rapid raise of the Euromarkets and the "recycling" of Petrodollars;
3. Rapid advances in information technology and communications (Cobb 1998, 7).

For Gorostiaga (1984, 2), offshore financial centers can be considered a "new articulation of world capital operating on a global market scale under the hegemony of finance capital," linked to the already mentioned characteristics of the current historical period. In the international division of labor, therefore, they can be considered as geographical points essential to reducing costs, and consequently increasing the profits of large companies and financial operators (Gorostiaga 1984).

However, the definition of "offshore financial centers" is not unanimously agreed upon, despite the recent accumulation of proposals on the subject, especially in geography and economics. According to Goldfinger (1986, 118), offshore financial centers are centers of international transactions with the following characteristics: 1. emission and reception points of global financial information; 2. receive and manipulate large quantities of this global "stateless money"; and 3. for this, offer a wide range of financial services, some of them with great operational sophistication. Although onshore financial centers have always been "common" throughout the history of capitalism (cities of Genoa, Venice, Bruges, Lyon, Paris, Frankfurt, London, etc.), offshore financial centers only came to be in 1960s and 1970s, at the very wake of the formation of the Euromarket, and the diffusion of information technologies (Warf 1989). Offshore financial jurisdictions in this sense are "merely the geographical extension of the Eurocurrency market outside Western Europe" (Zeromé 2007, 24), and as Wainwright (2017) states, they are no longer just the "corners" of the world economy.

One of the pioneers of the study of offshore financial centers in geography, Roberts (1994) points out that the reason for offshore financial centers stems mainly from their lax regulation of financial resources entering and leaving their jurisdictions. For the author, "they are essential moments in the spectacularly fast and free-wheeling world financial system, allowing institutions to cope with volatility and 'manage' varieties of risk facing international capital" (Roberts 1994, 92/93). This point was also noted by Machado (1996), who found for Brazilian businesses use of these regulatory differences between jurisdictions, with transactions that allow profits often exponentially higher for operators in offshore centers.

These arguments are also noted by Hudson (1999, 145), who affirmed that the two structural elements which allowed offshore centers are financial regulation and deregulation (liberalization of financial markets) and new information technologies. According to the author (Hudson 1999, 142), offshore financial centers play a key role in the breakdown of countries' sovereignty over finance (i.e., property rights and the regulation of processes in a territory). They are both an expression of the new global finance and a central element in the transition from embedded to disembedded economies. Like the Euromarket, OFCs have developed largely as an "alternative" to (or attempt to "escape") the regulation of traditional onshore financial systems, the national financial markets of the world's largest economies, which are heavily regulated (despite national variations in laws). Given the importance of the regulations that affect them, Choi et al. (1986) demonstrate that offshore centers can be considered "legal fictions."

These centers are part of a strongly interconnected network, which has an extremely hierarchical structure (Dollfus 1992). In most cases, new centers have more links with other "traditional" international financial centers (notably the three financial centers of the Triad: New York, London, and Tokyo), than with the local/national countries that harbor them, as also explained by Laulajainen (2002, 32). Although effectively operated in New York and London, newly established firms and/or bank accounts are mostly domiciled in Caribbean offshore jurisdictions (mainly in the Cayman Islands). This is the first essential characteristic of offshore financial centers: they are either competitors or complementary to traditional financial centers. They are *complementary* in that they carry out operations demanded by financial agents established in global cities, because they currently cover all the continental time zones, from Tokyo to San Francisco, through Hong Kong, Singapore, Bahrain, London, the Caribbean, New York, and Chicago. Because they are distributed throughout entire range of the different successive schedules, they allow international business to be carried out endlessly on a continuous basis. This complementarity nature is particularly evident in the interbank and foreign exchange transactions of large banks such as Citibank, HSBC, JP Morgan/Chase, Bank of America, Deutsche Bank, Barclays, BNP, etc., which since the 1980s have been present in practically all these centers and time zones (Goldfinger 1986, 119). On the other hand, offshore financial centers are *competitors* of global cities when they seek to attract new banks and supplementary funds, creating new markets, producing new financial instruments, and offering advantageous conditions for investments of international operators. This competition, as mentioned, is especially strong between centers in different time zones (such as between Hong Kong and Singapore, or even London and Luxembourg) (Goldfinger 1986, 119).

Competition between these different locations has become so acute over the last 20 years that even onshore economies have made efforts to attract investment and firms that traditionally would go to the already "traditional" offshore centers. This is the case, for example, of Singapore, Mauritius, and Cyprus, as well as Ireland, which has recently made extremely "generous" agreements with global companies, such as Apple, Amazon, and Google, to allow a large set of corporate operations—mostly involving financial transactions and/or payment of taxes—to be carried out there (Wójcik 2012a, 338). Even countries such as the US and Japan have been more "tolerant" in their local legislations to allow banks and financial operators to operate their balance sheets with transactions that are not taxed and regulated by conventional national legislation (Wójcik 2012a, 338). In the US, the state of Delaware, located between Washington DC and New York City, specializes in hosting institutions that carry out various types of financial operations, which had been typical only in offshore jurisdictions (Wójcik 2012a, 338).

However, the most perverse consequence of the development of offshore jurisdictions is related to the fiscal question. Most of the transnational corporations, whether financial or not, have systematically used these jurisdictions as places to formally register activities and operations, in order to avoid paying taxes in their countries of origin, which leads to two serious problems for any economy: tax avoidance and fraud or tax evasion. As Aalbers (2017) notes, the difficulties in distinguishing between these two practices further complicate the possibility of combating them. For Wójcik (2012a, 340), large firms open subsidiaries in offshore centers primarily aiming at legal registration of annual profits, or large financial transactions, which are minimally—or not—taxed in these offshore jurisdictions.

In addition to the tax avoidance and evasion that these centers allow, they also offer financial operators a very advantageous legislation in terms of the secrecy of information, both with the necessary documentation and information about the real owners of the established firms. A significant part of the activities carried out in these financial centers are "shell companies", since most firms only exist as accounting phenomena (Goldfinger 1986, 120).

Roberts (1994) observed for the Cayman Islands in the early 1990s, US$ 250 billion in bank liabilities, and 546 installed banks (banks from all over the world). Analysis of the urban landscape of its capital city, Georgetown, could only find 69 banks that had a "physical" agency in the city, which did not occupy more than five or six streets where it was possible to identify and actually visit these banks (Roberts 1994, 92). The other banks could be found only in the "brass of plastic name plate in the lobby of another bank, as a folder in a filing cabinet or an entry in a computer system" (Roberts 1994, 92). Instead of "real entities," they were mostly "paper entities."

In addition to allowing these operators to "escape" the regulations that are very expensive in their countries of origin (such as taxes, compulsory deposits, obligatory reserves, etc.), offshore centers function as "profit centers" to move profits of large firms—or family and personal fortunes—according to the best "temporality" or opportunity for their repatriation (Dicken 2010, 437).

Although situations such as these take place on several territorial scales, Peter Dicken (2010, 437) asserts that offshore financial centers are usually located in small territories—islands or microstates—and function as points of manipulation for the "Fictitious capital circuit" (Dicken 2010, 437). The reasons why offshore financial centers are generally territorial enclaves, or islands, of reduced geographic size are basically (Wójcik 2012a, 337):

1. They are small economies, it is more feasible for them not to charge or to charge very low tax rates; the same reasoning applies to the fees charged. If they were complex economies, their demands for fiscal resources would be greater; and
2. They are centers of financial operations that lodge less sophisticated operations in generating the value chain of financial globalization, activities are more computerized, and the amount of skilled labor required is also lower; therefore, the labor markets can also be smaller in size for offshore operations.

Contemplating this territoriality of the offshore financial centers, Machado (2017) proposes a classification with three divisions for them:

1. Former colonies and/or protectorates of former European empires such as the Cayman Islands, British Virgin Islands, Bermuda, Gibraltar, Anguilla, Guernsey, Isle of Man, Jersey in the Channel (Commonwealth territories or British Protectorates), Netherlands Antilles, and Aruba (protectorates of the Netherlands);

2. Small states and microstates specializing in various types of offshore financial services that provide their services to non-residents (banks and other financial units, corporations, law firms, individuals, etc.);
3. Consolidated territorial states that develop a series of regulations "to escape and differentiate regulatory and/or fiscal rules in some location or locations within the national territory" (Machado 2017, 332).

Under these conditions, it is possible to say that the largest offshore financial center in the world is the City of London, a part of the city has a more flexible regulation that allows non-resident actors to do business in currency other than the British Pound (mainly the dollar). This "model" spread throughout all English possessions, creating the offshore centers of the first kind described above. As a result, the UK reformulated the structures of its former empire, now under the command of finance (Machado 2017).

Another important characteristic of these centers is the historic-structural character of the international networks of which they are part. In other words, offshore financial centers are created from an inheritance of the international division of labor during the colonial period of world history. This heritage is a powerful element of the formation of offshore centers for two main reasons:

1. The (ex)metropolises are guarantors of military protection, and therefore of the political stability (internal and external) of these jurisdictions; and
2. The colonial heritage always means a certain linguistic homogeneity, and in most cases a cultural proximity and juridical institutionalization similarity. These resemblances are also fundamental for the fluidity of communications between offshore centers and global cities (ex-metropolises) (Wójcik 2012a, 337).

According to these definitions, it is possible to arrive at an approximate list of those parts of the terrestrial globe that can be considered offshore jurisdictions. This list of offshore jurisdictions was compiled using information from multilateral agencies, financial authorities, and scholars of the subject, published since 1977. According to Wójcik (2012a), defining what offshore jurisdictions are is not an easy task, since several criteria that are used, for different purposes, to identify this type of phenomenon, which is thus more a matter of "grade" than of "fact." The main elements that define this "degree" that allows the identification of offshore centers are 1. taxation elements, 2. regulation, and 3. secrecy of operations. From these criteria, the jurisdictions that can be considered offshore are listed in Table 1.1.

As can be seen from the above list, countries with a large territorial dimension and global economic importance—such as the United States and the United Kingdom—can also be considered offshore jurisdictions. In the case of these larger countries, of course, only tiny portions of their territories function as legal enclaves, as is the case of City of London in the UK and the state of Delaware in the USA.

Another major issue involving tax havens and offshore jurisdictions is their definitive contribution to the so-called "shadow banking system," currently "largely incorporated in offshore financial centers," through the creation of Special Purpose Vehicles (SPV) that are widely produced and marketed in these centers (Rixen 2013,

Table 1.1 Main offshore financial jurisdictions in the world (from Wójcik 2012a)

Countries	Number of citations
Bahamas, Bermuda, Cayman Islands, Guernsey, Jersey, Malta, Panama	11
Barbados, British Virgin Islands, Cyprus, Isle of Man, Liechtenstein, Netherlands Antilles, Vanuatu	10
Gibraltar, Hong Kong, Singapore, St. Vincent & Grenadines, Switzerland, Turks and Caicos Islands	9
Antigua and Barbuda, Belize, Cook Islands, Grenada, Ireland, Luxembourg, Monaco, Nauru, St. Kitts and Nevis	8
Andorra, Anguilla, Bahrain, Costa Rica, Marshall Islands, Mauritius, St Lucia	7
Aruba, Dominica, Liberia, Samoa, Seychelles	6
Lebanon, Niue	5
Macau, Labuan (Malaysia), Montserrat	4
Maldives, United Kingdom	3
Brunei, Dubai, Hungary, Israel, Latvia, Madeira, Netherlands, Philippines, South Africa, Tonga, Uruguay, US Virgin Islands, USA	2
Alderney, Anjouan, Belgium, Botswana, Campione d'Italia, Egypt, France, Germany, Guatemala, Honduras, Iceland, Indonesia, Ingushetia, Jordan, Marianas, Melilla, Myanmar, Nigeria, Palau, Puerto Rico, Russia, San Marino, Sao Tome and Principe, Sark, Somalia, Sri Lanka, Taipei, Trieste, Turkish Republic of Northern Cyprus, Ukraine	1

Source Wójcik (2012a, 337)

435). As Wójcik (2012a, 339/340) also affirms, there are several institutions, both financial and non-financial (from banks, through investment funds and holdings), that operate a significant part of their operations through these jurisdictions, with assets each time more complex. This was the case of Madoff Investment Securities (a famous North American fortune-making company, bankrupt in 2008 due to numerous frauds), which also operated a Bermuda-based risk management and control office. A similar procedure took place at the Northern Rock bank, which used the Granite, a registered trust in the Jersey Islands, to securitize the UK's most risky mortgages and invest in US subprime mortgages. Bear Stearns and Lehman Brothers had hedge funds primarily in Ireland and the Cayman Islands. Ábaco, another investment company investigated for fraud by the SEC, was registered in Delaware. The largest corporate scandal in the Western world has also been linked to these possibilities opened by the OFCs: Enron, a US energy company, operated through a global network of 3,500 domestic and foreign subsidiaries, including 441 registered in the Cayman Islands. Through these subsidiaries, highly complex tax evasion schemes were developed, which in turn were designed by three types of large global firms: 1. large consultancies, such as Arthur Andersen and Deloitte & Touche; 2. international banks, such as Chase Manhattan, Deutsche Bank, and Bankers Trust; and 3. large law firms specializing in global financial transactions. The scheme was able to leverage Enron's profits at first, but mainly increased the earning of its directors and CEOs. In a short time, however, the situation proved to be unsustainable, and the losses calculated with the bankruptcy of the company in 2005 were somewhere around US$ 5–10 billion (Wójcik 2012a, 340/341). Although offshore financial centers reduce the "friction" in the circulation of these resources, this lack of accurate accounting, the tax evasion, and fraud, as well as the promiscuous proximity of these circuits to money laundering of illicit activities is a permanent source of instability in the global financial system (Roberts 1994, 93). In addition to this instability, some of these centers have also facilitated illicit operations, particularly those related to money laundering by criminal organizations and drug trafficking (Hudson 1999, p. 144).[12]

Offshore centers in Latin America

The emergence of offshore centers in Latin America has mainly occurred in Central America and the Caribbean and has some factors that are common to all places of this type, mainly because of the "advantages" they present in relation to the "onshore" financial markets as well as by the less restrictions and controls on the financial activities carried out there (Warf 1989, 265).

For Hudson (1999, 143), these centers that emerged in the Caribbean took advantage of the higher costs involved in other financial markets on the planet (especially in comparison to the Euromarket in London), as well as the "diseconomies" that these markets—usually metropolitan markets—which have high costs for office/building rentals, high prices for necessary skilled labor, and less lean/efficient structures for companies. Secondly, they are in the same time zone as New York, which also provides advantages for the products they could offer (compared to the same types of operation in US market). These offshore centers offer the typical advantages of such areas with more permissive regulation, such as absence of exchange control, minimum deposits or reserves, interest rates, etc. Between 1965 and 1975, the branches of banks opened outside the US (in offshore centers) jumped from 180 to 732; in the Caribbean, this increase was from 5 to 164 (Hudson 1999, 43).

Gorostiaga (1984) made a fundamental contribution to the understanding of offshore centers in Latin America in his study of Panama in the international division of labor in the 1960s and 1970s. The basis for the formation of an international financial center in the country was derived from capitalist dynamics at the time, with the main causes being

1. Expansion of the Transnational Banks (TNBs), following and/or preceding their main clients (the transnational corporations);
2. The problem of the United States balance of payments in the mid-60s, "which induced change in credit policies and in banking laws and practices";

[12]Given its increasing importance for international financial flows, since the 1990s, the OECD has initiated a move to identify and demand changes in the permissive laws of offshore centers, including mainly greater requirements for openness and transparency of operations and reduction of preferential treatment provided to foreigners (Dicken 2010, 437). The pressure to regulate the OFCs increased after the September 11, 2001 attacks as part of the US efforts to control global money laundering (Dicken 2010, 438).

3. International monetary crisis, instability of the money market, currency fluctuations and concomitant speculation, mainly due to the creation of a large sum of liquid assets with Eurodollars and Petrodollars.

These elements are important in the formation of several new offshore centers, including mainly Hong Kong, Singapore, Bahamas, Beirut, and Panama. All these have already mentioned common features: they were considered tax havens, places for the development of paper companies (or "special purposed entities"), but also places for development of companies considered "Flags of convenience," as well as free trade zones ("Free zones"). In the case of Panama, its importance as a financial center was so significant that since the 1960s the maritime channel—the country's main economic infrastructure—became economically less important than the financial activities carried out in the country.

According to Warf (2002), the construction of the Panama Canal in 1903 is one of the central factors of the country's subsequent financial development. In addition to making that territory a central point for the international movement of goods, it also accommodated shipping companies called "flags of convenience," due to the tolerance of local legislation in relation to the development of these activities in the country. In a pioneering study, Johnson (1976) argues that Panama had rapid financial growth in the 1960s due to three initial factors: 1. its favorable geographical position (close to both the United States, the world's largest economy, and the other countries of America Latina); 2. its dollarized economy (which facilitated international business); and 3. the liberalization of financial laws, which increased the country's attractiveness to companies in the region (mainly USA, Colombia, and Ecuador). These arguments are also supported by Warf (2002), who points out that, in addition to the dollar, the local macroeconomy (interest rate, inflation, and exchange rate) was directly determined by FED policies, making the national business environment more conducive to installation of US financial operators. The liberalization of the economy began in 1970, with the action of finance minister Nicholas Barletta. Under the Carter–Portillo agreement, the new legislation liberalized all rules for banking operations, including the possibility of full operator secrecy, as well abolishing any kind of exchange control. This legislation made the country more attractive than Switzerland, which in the 1970s was the country that hitherto guaranteed the secrecy of its depositors (Warf 2002, 38). However, the most important contribution of this new legislation was related to the tax issue, which was created to exempt foreign companies from paying national taxes, including taxes on profits, capital gains, deposits, savings, etc. (Warf, *op cit.*, 39).

Also dating from the 1970s, several Latin American and Caribbean countries transformed into offshore financial centers. Gorostiaga (1984, 28) highlights the importance of four major jurisdictions that would soon become important financial centers of this type: first the Bahamas and the Cayman Islands, and later Bermuda and the Netherlands Antilles.

The Bahamas became, as early as the 1960s, a platform for conducting international business, especially financial ones. Gorostiaga (1984, 28) stated that the operation of the "combination of fiscal and tax exemptions, accessibility, stability, and an administrative apparatus which together make the country an attractive centre, especially for US Banks." Since 1973, this situation was reinforced with Bahamas tax rate stability (when at the same time all rates were being increased in the main traditional IFCs). A figure for the year 1976 represents the importance of the growth of the Island for the North American businesses: in May 1976, there were more foreign loans from the US Banks registered in the Caribbean than in London (Gorostiaga 1984, 28).

The Enciclopedia Latinoamericana (Sader et al. 2006) defines "tax havens" as countries that fulfill these functions already described. It characterizes tax havens as colonial territories that adopted flexible policies which refer to three main types of issues: 1. taxation on transactions carried out (low or non-existent rates), 2. guarantee anonymity to resource owners, and 3. in the case of corporations, the secrecy of the shareholder composition of the company (Sader et al. 2006, 911).

Following these criteria, the Encyclopedia argues that the 16 offshore jurisdictions in Latin America and the Caribbean include Anguilla, Antigua and Barbuda, Aruba, Bahamas, Barbados, Belize, Cayman Islands, British Virgin Islands, Dominica, Grenada, Montserrat, Panama, St. Lucia, St. Christopher and Nevis, St. Vincent and the Grenadines, and Turks and Caicos. Table 1.2 shows in more detail the main characteristics of these territories.

The table is organized in terms of the importance of direct financial investments, according to United Nations Conference on Trade and Development (UNCTAD 2017). For these financial investments, two British protectorates, the British Virgin Islands and the Cayman Islands, are ranked first, followed by Panama and the Bahamas. The territories, which are more dynamic in terms of the financial investment flows, are ranked worst in the Financial Secrecy Index of the Tax Justice Network (FSI 2018). Practically all of them have direct relations with their former metropolis, which attests to the permanence of the uneven structure of the international division of labor inherited from the historical *longue durée*. Finally, there are also countries or jurisdictions with a small territorial and demographic dimension, the largest being Panama (75,515 km^2 and 4,098,587 inhabitants) and the smaller Anguilla (90 km^2 and 14,909 inhabitants) and Montserrat (101 km^2 and only 5,177 inhabitants).

The UNCTAD report in 2017 entitled *World Investment Report* shows that in the case of the British Virgin Islands and the Cayman Islands, despite their strong link with the countries of Northern Latin America, their flow network covers countries in other continents. In the period from 2010 to 2014, the main investment flows to the two islands came from the following countries: 1. Hong Kong with US$ 148 billion (33% of totals), 2. the United States with US$ 93 billion (21%), 3. Russia with US$ 77 billion (17%), 4. China with US$ 40 billion (10%), and 5. Brazil with US$ 23 billion (5% of totals); other countries together made up the remaining 14% of investment stocks (UNCTAD 2017).

Table 1.2 Latin America—Main characteristics of offshore centers (2017)

Country/territory	Population	Area (km^2)	GDP/per capita (US$)	Legal status/relationship with ex-metropolises	Financial secrecy index ranking	Financial flows—US$ million (2017)
British Virgin Islands	31,196	152.8	30,989	British Overseas Territory	16	38,358.08
Caymans Islands	62,000	240	64,973	British Overseas Territory	3	37,433.21
Panama	4,098,587	75,515	14,407	Republic	12	5,319.20
Bahamas	395,361	10,010	29,825	Monarchy/British Commonwealth	19	927.73
Barbados	285,719	430	16,804	Monarchy/British Commonwealth	48	286.16
St. Christopher and Nevis	55,345	261	16,865	Monarchy/British Commonwealth	63	126.71
Aruba	105,264	180	25,593	Monarchy/Dutch territory	68	87.65
Antigua and Barbuda	102,012	442	14,858	Monarchy/British Commonwealth	98	60.89
St. Lucia	178,844	539	8,003	British Commonwealth	110	92.39
St. Vincent and the Grenadines	109,897	389	7,236	Monarchy/British Commonwealth	111	87.06
Grenada	107,825	348.5	9,878	Monarchy/British Commonwealth	101	79.36
Belize	374,681	22,810	4,773	Monarchy/British Commonwealth	90	77.00
Anguilla	14,909	90	22,060	British Overseas Territory	56	62.38
Dominica	73,925	750	7,483	Republic	96	18.88
Montserrat	5,177	101	12,284	British Overseas Territory	112	6.09
Turks and Caicos	35,446	616	27,850	British Overseas Territory	87	–

Source Sader et al. (2006); Financial Secrecy Index (2018); UNCTAD (2017) https://unctadstat.unctad.org/CountryProfile/GeneralProfile/en-GB/004/index.html

In addition to official statistics, recently the disclosure of secret data from a series of documents of the so-called *Panama Papers* shed light on the networks of investment flows among Latin American countries.[13] As shown by the Tax Justice Network (FSI 2018), in April 2016, these Panama Papers revealed to the public about 11.5

[13]The main investigation that systematized and made this data public was coordinated by the International Consortium of Investigative Journalists (ICIJ 2019). The consortium gathered data on some 785,000 offshore entities, which in turn included four investigations: the Panama Papers, the Offshore Leaks, the Bahamas Leaks, and the Paradise Papers.

million documents from the legal services firm Mossack Fonseca, which contained the main elements of the *modus operandi* of offshore investments made in Panama, as well as several other jurisdictions in Latin America and worldwide. According to the analysis of the Tax Justice Network (FSI 2018), the documents "showed the world what a few observers had long been saying: that the secrecy available in Panama makes it one of the world's top money laundering locations."

The website of the International Consortium of Investigative Journalists (ICIJ 2019) states that the Panama Papers reveal a series of accounts with varying degrees of irregularity for several large global companies (such as UBS and HSBC Bank), and at least 12 world heads of state (including the former president of Argentina, Mr. Mauricio Macri). The analysis of the Brazilian papers revealed in the scandal verifies that by 2016 there were 130,016 "unidentified" Brazilian offshore entities opened by Mossack Fonseca—in which it was not possible to define their locations more precisely—and another 1,432 entities were identified, and their distribution in the world was as follows: Panama (521 entities), British Virgin Islands (473 entities), Nevada, USA (132), Niue (91), Seychelles (79), Samoa (43), Bahamas (33), Cayman Islands (21), Anguilla (14) United Kingdom (6), Bermuda (4), Wyoming (4), Isle of Man (3), Barbados (2), Hong Kong (2), New Zealand (1), and Uruguay (1) (ICIJ 2019).[14]

The international business and law firm Mossack and Fonseca, which began operations in 1977, by 2016 had representation offices in 39 countries, including Brazil. In addition to facilitating operations in Panama, the company also offered its services so that interested customers could conduct operations in 21 other jurisdictions, most of them offshore financial centers; these centers included jurisdictions within the USA, UK, and Hong Kong, but primarily the British Virgin Islands, Isle of Man, Jersey Islands, Bahamas, Uruguay, Cyprus, Malta, and Seychelles (ICIJ 2019). The main offshore center in which Mossack Fonseca have opened companies for its clients was the British Virgin Islands, with almost 113,000 firms, almost twice as many as in Panama itself, with about 48,000 firms opened. Following these two centers, the Bahamas and Seychelles are the other two most sought "destinations" for Mossack's customers (ICIJ 2019).

The service sold to open an account and a company in one of these centers is a relatively simple process and can be divided into four main steps: 1. the Mossack Fonseca office is sought by individuals or companies (mainly financial firms, including banks or law firms); 2. according to the degree of confidentiality required by the client, Mossack Fonseca offered a "menu" of countries, so that clients can choose the one that best fit their "needs"; different jurisdictions, depending on its internal regulation, could be more suitable for each client's situation; 3. the company also offered "shelf companies," that is, companies already formed and ready to be "employed" to move resources; 4. in establishing or buying a company, the great advantage to the investor is that their identity is hidden. This opened company, in turn, becomes a

[14]Data from the International Consortium of Investigative Journalists about Brazil can be found at the following address: https://offshoreleaks.icij.org/search?utf8=%E2%9C%93&q=&c=BRA&e= (Access on 19.02.2019).

global "player" with its own legal identity and is able to open bank accounts, acquire assets, and make investments in any part of the world (Estado de São Paulo 2016). As the report further shows, Mossack and Fonseca

> is one of the largest creators of front companies, corporate structures that can be used to hide assets. The internal archives of the law firm contain information on 214,488 offshore organizations linked to people from more than 200 countries and territories" (Estado de São Paulo 2016).

1.4 The Formation of International Financial Centers and the Relative Position of São Paulo

The reorganization of main capital accumulation centers and financial institutions followed closely the diffusion of these infrastructures of information. If on one hand the productive economy became globalized with certain degree of deconcentration (alongside the relocation of production units of multinationals in peripheral countries) (Fröbel et al. 1980; Gottdiener 1989), the financial control of new business networks created became even more circumscribed, and their localization has privileged "international financial centers." How to define this type of cities? What is the role of São Paulo as a financial center in the current world system?

The concept of international financial center has several definitions, mainly in geography, but also in economy and in sociology. According to a general proposal from Laulajainen (2002, 332):

> the ability to collect, exchange, rearrange and interpret information is the most persistent characteristic of an international financial center. Collection depends largely on factors external to the center: convenient river crossings and economic shipping routes in the early days; railroad hubs, ground cables and motorways later on; and flight connections and satellite communication today

The definition of Laulajainen shows the importance of transportation and information transmission infrastructures as one of the main conditions for the development of international financial centers. It is a real consensus that "information" is in the very basis of the formation of these centers, as shown by several authors, like Thrift (1994), Porteous (1995, 1999), Martin (1999), Santos (2000), and Wójcik et al. (2018).

In addition of drawing attention to the importance of the factors related to this technical-scientific-information milieu in the current operation of financial centers, it is also important to highlight that the development of these centers is a result of geohistorical processes slowly built, many times along centuries (as shown in items 1.1 and 1.2). It is exactly for being dependent on the combination of "rare" factors, such as sophisticated infrastructures, firms, and specialized human resources, that the financial localizations are very dynamic, but also "always slow to change" (Demole *apud* Cassis 2006, p. xi).

The first analysis that incorporates the temporal dimension in the formation of international financial centers was made by economic historians, and almost all incorporate an evolutionary bias in their arguments. The work that inaugurates this approach is from the historian N. S. B. Gras, in *An Introduction to economic history* (1922). For the author, the formation of metropolis—in the case of the Western Europe economies—would be a process of four evolutive phases, the last being the "development of financial organization." The evolution of the division of labor starts with the development of retail trade activities (phase 1), passes through industrial development (phase 2), afterward through transportation enhancement (phase 3), culminating in the financial development of metropolis (phase 4). For the author, "It is to be expected from the nature of things that the metropolis will develop its financial resources to the limit […]. As the general organization of the market is completed, the manufactures develop, and transportations facilities built up, a metropolitan financial system is constructed" (Gras 1922, 243).

The first book systematically devoted to the historical analysis of the international financial centers was written by Kindleberger (1974), entitled *The Formation of Financial Centers: A Study in Comparative Economic History*. The evolution of financial operators in these centers, according to Kindleberger (1974, p. 9), would meet the following steps: 1. banking activities started out to serve the needs of sovereigns and nobles; 2. further development was due to the connection with commerce activities (mainly international trade); 3. then financial activities begun to be related with State/governmental finance; 4. one next step was the proximity of banks with the financing of transport activities, including shipping, canals, turnpikes, and railroads; 5. financial operators started also to finance industry and large-scale productive activities; 6. and finally these actors engaged with intermediation in insurance, mortgages, consumer finance, factoring, pension funds, and the like.

The fact is that in all countries and cities analyzed by the author (England/London, France/Paris, Germany/Berlin, Italy/Rome and Milan, Switzerland/Zurich and Geneva, USA/New York and Canada/Montreal and Toronto), elements from local physical geography and factors associated to transportation basis (i.e., the relative localization of the city in a circulation network previously existing) were fundamental for the success or failure of these cities. The internationalization of the activities, on its turn, is responsible for the acceleration of the main mechanism that strengthens the financial centers: the generation of economies of scale. For Kindleberger (1974, 11):

> Up to a certain high degree of concentration, positive externalities and economies of scale appear to outweigh diseconomies, favoring centralization. The continuous reduction in the costs and difficulties of transport and communication over the last two hundred years has favored the formation of a single world financial market

The attraction of companies is made mostly because of the face-to-face contact that is allowed in these densely urbanized areas. These contacts accelerate the transmission of information in these places, providing for more rational decision to be made, and reducing, therefore, the uncertainty of economic processes. These "advantages" are typical of a great market that is formed from large cities that offer greater liquidity, safety, and speed in businesses performed therein (Kindleberger 1974, 10). According to the author, these economies of scale are such a powerful factor that leads to the concentration of activities in a single financial center dominant within national spaces, and that also tends to result in the emergence of a single world financial center with highly specialized functions (mainly international loans and clearing payment chambers between countries). This concentration, on its turn, tends to be a self-reinforcing process: banks, investors, brokers, and insurers—among other businesses of financial character—establish branch offices in these centers, what catalyzes their growth, specialization, and centrality.

For Reed (1981, 57 and ss.), the development of financial centers involves five main phases. In the first one, the center serves only the immediate surrounding with banking services; in the second phase, this operation area is enlarged, reaching other cities, but in a still regional scope; in a third phase, the financial center becomes predominant in all national territories in which it is part. In the fourth phase, the center diversifies its financial activities (offering also other types of services, i.e., non-banking financial services), extrapolating the territory limits in which it is located. In the fifth and last phase, the financial center becomes the apex of the urban network hierarchy from which it is part, offering the broadest range of services and financial products to their customers. In the author's definition:

> A financial center is an urban area which contains a concentration of specialized institutions that possess, at least marginally, the international skills and capabilities necessary to facilitate the flow of goods, services, information, and capital between its own national economy and the other national economies of the world (Reed 1981, 58).

One of the most important recent analyses of the functioning of international financial centers is that of Roberts (2000). In the author's view, the main factor that allows the emergence of these types of centers is the "agglomeration economies" generated by the concentration of "complementary activities" that exist in these cities. As financial institutions are large demanders of specialized services of advocacy, accounting, information technology specialists, consultancies, audits, the existence of these related firms is an essential development factor of the international financial centers. With this reasoning, the author is also emphatic about the importance of information as a central element of financial dynamism: "above all, the crucial factor for the success of financial institutions is information. The quantity and quality of information is vital for the competitiveness of financial firms" (Roberts 2000, 126). In addition to this central role played by information, the author also lists the following factors as crucial for the development of international financial centers: 1. peace and political stability; 2. solid legal systems with "international trust"; 3. modern communications infrastructure (to enable the flow of information as efficiently as possible); 4. presence of skilled labor; 5. existence of dynamic financial markets; 6. development of complex financial instruments; 7. efficient supervisory and regulatory mechanisms; 8. "market economy," i.e., "freedom for companies and individuals to save, invest, work and consume"; and finally 9. modern general urban infrastructures (mainly transport systems to the fluidity of people and documents). For Roberts, international financial centers can be defined as:

> cities that specialize in providing international financial services. Its function is to intermediate financial flows between countries. This is done by the financial markets and the

> companies located in them. The set of financial institutions present in a financial center determines the range of activities developed there (Roberts 2000, 124).

Another important contribution of its proposal is a classification in relation to the four types of financial centers that exist nowadays: 1. Global financial centers: offer a wide range of financial products and services, especially wholesalers, to a clientele also of global companies. For Roberts, in the year 2000, there would be only three such centers in the world: New York, London, and Tokyo (each one occupying the relative center of the corresponding time zone); 2. International regional financial centers: even if the term is paradoxical, this second classification encompasses those centers that have an activity that goes beyond the national territory in which they are installed and provide services to different countries in the continents where they are located; 3. Offshore financial centers: they are more like "contact points" than as centers of accumulation of wealth and carry out operations that have little—or no relation—with the territories in which they are installed (as we saw in item 1.3); 4. National financial centers: they only serve the needs of the country concerned, working mainly in retail and bilateral trade services between the countries involved (Roberts 2000, 124/125).

A second set of authors that is in the base of the studies of international financial centers adopts a more critical vision of the phenomenon, using the concepts and approaches of political economy (mainly Marxist approaches) to perform their analysis (Bassens and Van Meeteren 2015). The three characteristics that unite these visions are as follows:

1. They consider that the development of financial centers in the early twentieth century is the result of a prior capital accumulation process, associated to dynamic "productive" activities (such as agriculture, commercial, and mainly industrial operations); financial activity does not come "out of nowhere";
2. This literature also treats financial activities within a social division of labor, including the ancillary services that gravitate around financial agents, the so-called "corporate" (Sassen 1991) or "business services" (Daniels 1993);
3. For their reasoning, the concepts of territorial division of labor and uneven development are crucial.

The main author that inaugurates this current is Cohen (1981), largely influenced by Stephen Hymer's classic study on *The Multinational Corporation and the Law of Uneven Development* ([1972] 1983). For Cohen, as from the end of the World War II, the former international division of labor (founded from the relation of colonies that exported raw materials and imported finished industrialized products) was expressively modified, with the international spread of productive activities, capitalized by the investment of multinational firms. In the beginning, peripheral countries start to produce semifinished goods and important parts/components for production chains that start to operate in a division of labor with global amplitude (despite being intra-firm chains of production/trade). This new logic of expansion beyond the rich countries from the center of the world system is enlarged by the advantages that these conglomerates find in the peripheral countries, countries that offered cheaper—and less unionized—labor force, absence of environmental regulations, free trade zones, tariff exemptions, and joint ventures with local companies (Cohen 1981, 289).

The multinational corporations, therefore, start to deal with a system of branches/subsidiaries much more dispersed and diversified, and the control over financial information between them becomes more strategical. Together with these Foreign Direct Investments (FDIs), there was also a modification in the distribution of productive activities, due to the fact that these investments led to an international spread of demand for international corporate-related services. This process tended to increase the necessity in these same peripheral countries of multinational banks, specialized law firms, consultancy and advertisement companies in addition to "contacting firms" (Cohen 1981, 293). Also shown by Simon (1984, p. 89), "the multinational firms need for their own operations a bank that is present in the same countries they are."

This new international division of labor is also directly related to the rise of an international capital market system, less subject to national regulations and its respective central banks (and therefore a market more connected to necessities of transnational companies). This phenomenon explains the increasing importance of the international financial centers in countries from the core area of global capitalism, mainly New York, London, Frankfurt, and Zurich. In addition of being financial centers, these cities also shelter headquarters of multinational corporations. The corporative decisions made in these centers have, therefore, influence in all other cities/countries directly connected to the productive chains of these multinationals. According to this international division of labor that is installed, the financial centers that were created in the Third World in this period were only "appendices" of the international system of financial markets, and the more important units operate much more as "tax heavens," such as Panama, Bahrein, Singapore, and Hong Kong (Cohen 1981, 293). These cities serve "primarily as centers for moving and mobilizing financial resources, rather than centers of control" (Cohen *op. cit.*, 308).

Friedmann (1986), when developed his analysis about the "world cities," elected the finance variable as one of the main elements of its constitution. Finance agents and facilities are those that allow a more precise definition of the relative position of cities in the international division of labor. For the author, the world cities are those that became "basing points for global capital" (Friedmann 1986, 69) and are the result of a "spatial organization of the new international division of labor." Among these functions that this new international division of labor attributes for world cities, he emphasizes 1. headquarter functions, 2. financial functions, and 3. the articulation of regional and/or national economies with the global economic system.

The main elements that define what the world cities are, and that allows him to suggest their hierarchy, follow these characteristics:

1. Major financial centers;
2. Headquarters for transnational corporations (TNCs), including regional headquarters;
3. International/multilateral institutions;
4. Rapid growth of business services sector;
5. Important manufacturing centers;
6. Major transportation nodes;
7. Population size (Friedmann 1986, 72).

Using this methodology, the hierarchy of world cities that he defined for the beginning of the 1980 decade is shown below. It is important to observe that in Friedmann's analysis, São Paulo plays an important role as a world city of primary order, in the set of countries of the so-called semi-periphery of capitalism (Table 1.3).

Sassen (1991, 1993, 1999) is another important author that provides a more critical vision to the formation of what she defined as "global cities" (1991) or "global financial centers" (1999). These cities are on the basis of the "alteration of world economy structure" which occurred from the 1960 decade. According to Sassen, the restructuring of world economy resulted in a new international division of labor, with the following main transformations: 1. dismantling former industrial centers in USA and England; 2. accelerated industrialization in several Third World countries (especially in Asia, but also in big economies of Latin America); and 3. rapid internationalization of financial industry (most of all as from the 1980 decade).

The main thesis of the author is that since the 1960' occurred a combination of spatial dispersion of productive units/activities with the correspondent concentration/centralization of the control of these activities, increasing, therefore, the importance of the "global cities." These cities concentrate greater number of headquarters

Table 1.3 The World City Hierarchy according to Friedmann (1986)

Core countries		Semi-peripheral countries	
Primary	Secondary	Primary	Secondary
London	Brussels	São Paulo	Johannesburg
Paris	Milan	Singapore	Buenos Aires
Rotterdam	Vienna		Rio de Janeiro
Frankfurt	Madrid		Caracas
Zurich	Toronto		México City
New York	Miami		Hong Kong
Chicago	Houston		Taipei
Los Angeles	San Francisco		Manila
Tokyo	Sydney		Bangkok
			Seoul

Source Friedmann (1986), 72

of three types of companies: 1. financial companies (banks, stock exchanges, insurers, credit card companies, etc.); 2. corporative services (or advanced producer services), i.e., firms that provide "informational inputs" for decision-making of other companies (such as consultancies, auditors, publicity and marketing companies, accountability and law firms); and 3. industries with high technological capacity. In the author's own words, "the more a firm disperses operations, the more complex and centralized its top-level management operations become" (Sassen 1999, 78).

This new international division of labor, which has its epicenters in the global cities, produces a new "global elementary hierarchy of cities" (1993, 198), responsible for a power loss of States to control economic activities that occur in their territories. This global hierarchy would have three leader cities in financial terms—London, New York, and Tokyo—and other two main ones after (Frankfurt and Hong Kong) (Sassen 1999). For the author, "New York, London and Tokyo have disproportionate concentrations of top-level headquarters in the financial, industrial, commercial and producer service sectors," what places them in the top of world urban hierarchy (Sassen 1991, 169).

The main historical mechanism that gave rise to the global financial centers for Sassen is the process prior to the national consolidation of financial systems, which appeared in all countries in a city with greater concentration of financial firms. These cities start to control flows and activities that were spread through their territories and became, therefore, powerful capital accumulation centers. Historically, it is mentioned the case of New York, which overcame in financial terms all US cities. She also reminds other cases in countries more peripheral to financial globalization, such as Canada (where Toronto overtook Montreal), Australia (Sydney overtook Melbourne), India (Bombay became more important than New Delhi and Calcutta), and also Brazil, once São Paulo overtook Rio de Janeiro as the main national financial center (Sassen 1999, 76/77). All cities that are only indirectly connected to the main circuits of global financial transactions—i.e., all more "peripheral" cities—would be much more "gateway cities," agglomerations with importance for their urban networks and national economies, but that do not command the directions and the logic of global circulation of wealth.

Two other factors are essential in Sassen's explanation of global financial centers: 1. the dependence on the financial activities of information and 2. the consequences that arise from this for localization—or embedding—of these firms. For the author, the logic of information—especially financial information—obligates financial companies to be located in areas where there already is a concentration of extremely specialized professionals with sophisticated skills. The localization and daily face-to-face contacts of these actors allow them to share information and better understand the risks and opportunities in the unstable global financial market.

These "local" operation conditions of financial firms are, therefore, the thorough proof that even the financial market, that is the more "immaterial" and "footloose" of all markets, is conditioned by local dynamics, which are not economic and neither digital. It may be said, however, that "[…] these digitized markets are embedded in complex institutional settings" (Sassen 2005, 32), and more specifically there is a "locational and institutional embeddedness of the global capital market" (2005, 33).

The main manifestation of these fitting elements of financial markets is the proper local institutions (mainly the laws and codes that guide the market), which influence decisively in the good progress of financial businesses.

The only work in the literature about world cities and financial centers that directly treats Latin American metropolis is from Meyer (1986), entitled "The World System of Cities: Relations Between International Financial Metropolises and South American Cities." Strongly inspired by world-system theory of Immanuel Wallerstein, the argument of Meyer is very elucidating of some financial organization standards of South American metropolis that lasts until today. Under a more general point of view, its proposal starts also from the verification that the formation basis of global cities is the internationalization of multinationals occurred after World War II, which since the 1950 decade created "a network of international economic transactions within and between businesses such as industrial firms, banks, commodity firms, and trading companies." (Meyer 1986, 553). Because of this re-localization of large multinational firms, the world could be divided into three main types of areas, with its respective control metropolis:

1. core cities in older industrialized or postindustrial regions which contain most corporate headquarters;
2. semi-peripheral cities in rapidly industrializing areas;
3. peripheral cities in the poor nations of the world (Meyer, *op. cit.*, 558).

To understand this hierarchy of world urban network, therefore, it is necessary to provide special attention to the financial companies, mainly international banks. For Meyer (1986, 562):

> Finance is one of the key functions in the world system of cities, therefore an analysis of it provides a significant test of the ideas. Dominance is directly measured with the international bank headquarters-branch office (or subsidiary) link. International banks establish branch/subsidiary offices in cities to acquire and disperse capital and to provide other financial services. This branch/subsidiary office link, similar to the corporate headquarters-lower level unit link of previous studies, is a measure of control and coordination which embodies both dominance and force[15]

The South America countries—and adding also those from Central America and Caribbean—may be considered as belonging to the periphery and semi-periphery of world-system, and all structuring of its territories and internal economies result from this lower relative position in the international division of labor, as observed the main authors of Dependency Theory, such as dos Santos (1970), Amin (1974), Marini (1976), and also the World system theorist Wallerstein (1976). More recently, Panreiter (2015) states that Latin American cities appear only marginally in the map

[15]Still about the importance of headquarters, the author shows that they "focuses on long-run policy and strategy, acquires capital, interacts with corporate headquarters of other multinationals, and provides overall control and coordinative functions for a vast flow of goods, money, and information within the firm and with other firms. Each divisional headquarters implements the policies of the corporate headquarters and controls and coordinates the departments, while each of the latter focuses on one intermediary activity such as purchasing, sales, or accounting." (Meyer 1986, 557).

of the largest global corporations.[16] Arises from this relative position also the fact that the metropolises in the region are little integrated with each other, since historically their relations have always been stronger with large cities and headquarters of corporations installed in countries within the center of world-system. "South American metropolises have few international linkages within South America because peripheral cities are not highly integrated internationally with each other," reminds Meyer (1986, 564).

To analyze the relative position of South American metropolises, the author identifies the number and localization of international banks in the region, in order to create a hierarchy from them. For the year analyzed, 160 international banks were found in South American cities, where the three main cities that sited banks of this type were São Paulo, Caracas, and Buenos Aires. Together with the other three most important cities (Rio de Janeiro, Bogota, and Santiago), they held two-thirds of the total of foreign bank offices in the region (Meyer 1986, 569) (Table 1.4).

São Paulo already stood out at the time as the main financial center of South America, for having the greatest number of bank institutions in the region (as well as was the city of greatest populational weight). It is also interesting to observe that the three main financial centers of the region were São Paulo, Caracas, and Buenos Aires, significantly more important than the other centers—especially Rio de Janeiro, Bogota, and Santiago—in terms of number of existing international banks. In respect to Brazilian territory, the number of foreign banks in São Paulo was almost double from the city of Rio de Janeiro, with 54 offices—and almost 20 times more important

Table 1.4 South America—International bank offices in leading cities (1984)

City	Country	Population (×1000)	No. of banks	% of total
São Paulo	Brazil	8,408	94	17.8
Caracas	Venezuela	2,664	76	14.4
Buenos Aires	Argentina	9,910	71	13.5
Rio de Janeiro	Brazil	5,395	54	10.2
Bogota	Colombia	2,855	40	7.6
Santiago	Chile	3,692	24	4.6
Lima	Peru	3,303	16	3
Montevideo	Uruguay	1,173	13	2.5
Asuncion	Paraguay	602	13	2.5
Quito	Ecuador	743	11	2.1
La Paz	Bolivia	655	8	1.5
Totals		39,400	420	79.7

Source Meyer (1986, 570)

[16]For the author, South American companies generate together only 2.5% of the sales as sales profits as largest global companies, according to Forbes 2000 (Panreiter 2015, 9).

than the city of Brasília (that had only four office of International Banks) (Meyer, *op. cit.*, 569).

In Brazil, since 1980 decade some geographers proposed their interpretations to explain this primacy of São Paulo as the main national financial center. The pioneer study of Cordeiro (1986/87) shows how the concentration of commercial banks favored the higher position of São Paulo in the national urban network. Based mostly on the classic analysis of "major control points" of the American economy from Borchert (1978), the author shows how the "hypercephalic" banking concentration in São Paulo was stronger than the previous concentration of industrial firms, making the city the great point of "decision and command of transactional economy" in Brazilian territory. Correa (1989) analyzes what he denominated as "centers of territorial management" (*centros de gestão do território*), strongly based on an empiric analysis of headquarters and bank branches localization, seeking to identify the dialectic of "concentration and dispersion" of financial firms in the territory. The author shows that "The prominence of São Paulo arises from the fact that it concentrates a considerable part of head offices of Brazilian financial companies, among them the most important ones" (Correa 1989, 19), and this condition allows São Paulo to exercise a double role in the Brazilian urban network: 1. an effective territorial management center at the national level and 2. a gateway city (international management center) (CORREA, *op. cit.*, 20). Dias (1995), in her *Réseaux d'information et réseau urbain au Brésil* defines the straight relation of bank networks with other two main contemporaneous networks: the urban and telecommunication networks. These infrastructures are also heavily concentrated in the state of São Paulo, what helps us to understand how the metropolis of this region became the main national financial center.

Santos (1994a, b; 1996; 2001) produced some decisive interpretations about the consolidation of São Paulo as the main Brazilian financial center, in two books: *Por Uma Economia Política da Cidade* (SANTOS 1994b) and *Brasil: Sociedade e Território no Início do Século XXI*, written with María Laura Silveira (Santos and Silveira 2001). For the author, the primacy of São Paulo in the national urban network started to be built in the beginning of twentieth century, when the city becomes a dynamic industrial center and capital of a state with rapid development of agricultural circuits of production/exportation (mainly coffee production circuits).

As from the 1950 decade, São Paulo also turns into the locus of concentration of more dynamic non-industrial companies, especially those on commerce, services, and finance sectors. The most important advertising, consultancies, business firms, and all other information-intensive activities were installed there. São Paulo has since then a great number of universities—public and private ones—that make all types of professionals available for the companies, in several sectors demanded by local and regional economies. It was developed in this metropolitan region a complex of activities and infrastructures destined for the provision of services, such as airports, hotel and tourism facilities, cultural centers, which are at the same time functional for the localization and growth of dynamic companies in its urban area.

São Paulo becomes, therefore, an "international metropolis of Third World" (Santos 1994b), with all contradictions that arise from the fact of being a great city in a peripheral country: at the same time that hosts masses of poor and low-income population, it receives headquarters and central offices of the most dynamic companies, national or foreign, characterizing also what the author called the "upper circuit of urban economy," highly sophisticated and dynamic (Santos [1975] 1979). It is exactly this concentration of companies' headquarters and demands for corporative services, which grants to São Paulo the urban primacy that disposes in the Brazilian territory. For the author:

> Without ceasing to be the industrial metropolis of the Country, despite of the production deconcentration movement recently verified, São Paulo becomes, also, the metropolis of services, tertiary metropolis—or even better, quaternary—, the great center of decisions, the great factory of ideas that are transformed in information and messages, from which a considerable part are orders (Santos 1994b, 40)

How to measure, therefore, the recent importance of São Paulo as an international financial center? Which variables may be listed to show this primacy that the city gained in the last decades, both in national scale and in the Latin American context? Our analysis was based on the study of two main elements of the financialization of economies in the region, to identify the primacy of São Paulo as the largest Latin American international financial center: the study of stock exchanges and national banking systems. At the end of the chapter, we also show an analysis made by Z/Yen Group Limited, specialized in the construction of a semiannual document, which seeks to demonstrate which are the main international financial centers in terms of their "competitivity," since the year 2007.

As demonstrated by Wójcik (2011, 14), the importance of stock exchanges (and its undeniable geographical nature) arises from three main factors:

1. They are a powerful mechanism to reduce time and space constraints to firms that need financial resources (once they finance long-term projects, beyond what the local credit markets generally offer);
2. They diversify the funding sources of companies (which may reduce their dependencies in respect to banks);
3. By the innovative nature of products and financial services that the stock exchange promotes, they are important instruments for the diffusion of innovations in the countries in which they operate.

The historical development of the capital markets in Latin America occurred in a proper temporality, in the rhythm of the diffusion of capitalist mode of production toward the more peripheral countries and was modulated by structural obstacles in creating important financial centers, as Klagge and Zademach (2018) shows to the African case. If we think in a historic perspective, the stock exchanges in Latin America have a practically secular development. As described by Nabarro (2016), it is possible to make a periodization of the stock exchanges in the region, according to the foundation year: 1. some institutions are over 150 years old (such as the Stock Exchange of Buenos Aires and of Lima), which would form a set of countries that could be considered as "pioneer markets" in Latin America; 2. those that had their stock exchanges created in the twentieth century, up to the 1970 decade—therefore, before the "stock exchange revolution" (Beteille 1991)—and may be considered as "intermediary markets"; and finally 3. there is a set of countries in which the stock exchange has a much more recent existence (having been created as from the 1980 decade), which may be considered as countries of "late markets" (Table 1.5).

Even though some stock exchanges in the region are seculars, none of them reached the performance of their congeners from countries of the center of the world system, considering the structural differences of peripheral economies, when compared to these central countries. In the case of peripheral countries, some stock exchanges obtained a significant development, such as the those from BRICS (Brazil, Russia, China, and India), as observed Wójcik (2011), but are still part of economic systems in which the capital market does not have the dynamism of those in developed countries.

Table 1.5 Latin America—Evolution of the creation of Stock Exchanges (1851–1994)

Periodization	Creation year	Name	Country
Pioneer markets	1851	Rio de Janeiro Stock Exchange	Brasil
	1854	Buenos Aires Stock Exchange	Argentina
	1860	Lima Stock Exchange	Peru
	1867	Montevideu Stock Exchange	Uruguay
	1884	Guayaquil Stock Exchange	Ecuador
	1890	São Paulo Stock Exchange	Brasil
	1893	Santiago Stock Exchange	Chile
	1894	Mexico Stock Exchange	Mexico
Intermediary markets	1928	Bogota Stock Exchange	Colombia
	1947	Caracas Stock Exchange	Venezuela
	1970	National Stock Exchange	Costa Rica
	1977	Asuncion Stock Exchange	Paraguay
	1979	Bolivian Stock Exchange	Bolivia
Late markets	1987	National Stock Exchange	Guatemala
	1988	Santo Domingo Stock Exchange	Dominican Republica

(continued)

Table 1.5 (continued)

Periodization	Creation year	Name	Country
	1990	Panama Stock Exchange	Panama
	1991	Honduran Stock Exchange	Honduras
	1992	El Salvador Stock Exchange	El Salvador
	1994	Nicaragua Stock Exchange	Nicaragua

Sources Nabarro (2016), BM&F Bovespa (2017)

As shown by Carvalho et al. (2015), regarding their financial systems, the Latin American economies are mainly characterized by the following aspects:

1. The "financial depth" of countries in the region is limited, i.e., the relative participation of financial richness in the total richness generated is relatively low, especially when compared to the countries from the center of the world-system;
2. The financial systems are fundamentally based on banks, commercial and/or universal, mainly private ones (national or foreign), except for the Brazilian case (in which public banking system is very strong); due to this economic power of commercial banks (which became national oligopolies in all countries), the financial intermediation margins and profits are high (compared to international standards);
3. The capital market (securities, equities, as well as market for corporative debts) is limited and little liquid, especially if compared to the securitization industry found in USA and England.

Data from World Federation of Exchanges (https://www.world-exchanges.org) shows how this financialization of countries has recently occurred, in terms of development and stock exchange operation. Table 1.6 sought to emphasize the capitalization of the greatest stock exchanges in Americas, as well as the values of the most important exchanges in other continents.

It is possible to identify the little importance of the Latin American stock exchanges in respect to world stock exchange capitalization (which may also be seen in Map 1.1). When adding the values moved only by the Latin American stock exchanges listed by World Federation of Exchanges (US$ 1,998,718.9), we see that they represented in 2017 only 2.4% of world total. The two greatest stock exchanges in the region, the B3 (São Paulo) and the Mexican Stock Exchange, transacted, respectively, 1.1% and 0.5% of world total. Still in the context of Americas, the relevance is greater, but much lower than the importance of North American Stock Exchange. The New York Stock Exchange is around 23 times greater than B3 and 53 times greater than the Mexican Stock Exchange. Among the Latin American economies, B3 represented—practically in every year—almost half the movement, what demonstrates its importance for the region as a whole.

Table 1.6 Major stock exchanges in the world—Domestic Market Capitalization (US$) (2014–2017)

Year	2014	2015	2016	2017
Exchange				
Americas—total	30,269,000.0	28,034,457.8	30,918.621.2	36,486,553.1
NYSE Group	19,351,417.2	17,786,787.4	19,573.073.6	22,081,367.0
Nasdaq—US	6,979,171.9	7,280,752.1	7,779.127.0	10,039,335.6
TMX Group	2,093,696.7	1,642,516.7	1,993.522.7	2,367,131.6
B3 (Brasil Bolsa Balcão)—São Paulo	843,894.1	587,925.9	758,565.9	954,711.1
Mexican Stock Exchange	480,245.3	335,679.3	350,809.5	417,021.6
Santiago Stock Exchange	233,245.4	204,591.7	212,479.7	294,675.8
Colombia Stock Exchange	146,745.6	90,915.3	103,818.6	121,477.1
Lima Stock Exchange	78,839.8	57,560.1	81,088.9	99,218.6
Buenos Aires Stock Exchange	60,142.0	45,8789	63,614.3	108,739.9
Bermuda Stock Exchange	1,601.4	1,850.0	2,520.6	2,874.8
Asia/Pacific—total	21,081,883.1	24,102,415.8	24,281,463.2	31,296,841.7
Europe/Middle East/Africa—total	12,145,148.1	11,414,262.3	12,003,168.0	14,728,745.4
WFE total	63,496,031.3	63,551,136.0	67,203,252.5	82,512,140.2

Source World Federation of Exchanges (several years)

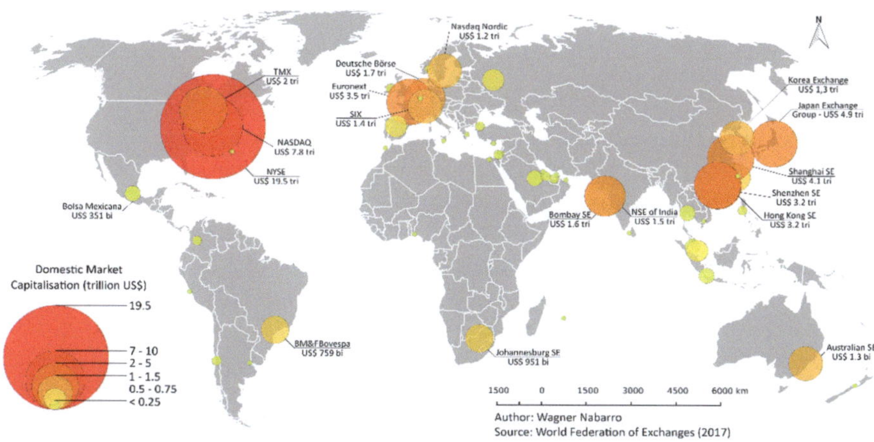

Map 1.1 The main stock exchanges of the world according to WFE (2017)

If the relative position of São Paulo, and the metropolises in Latin America, is frankly unfavorable in the international division of financial labor, what may be said in terms of the banking systems in the region?

As mentioned above, the literature about international financial centers is almost unanimous in emphasizing the importance of banks (be they commercial or investment banks) for the definition of the financial importance of an urban agglomeration (Labasse 1974; Cassis 2006; Chen and Chen 2015). In three different articles, Choi et al. (1986, 1996, 2003) show the significance of banks as essential elements for the formation and performance of international financial centers. To define the hierarchy of these centers, the authors work with the statistics of headquarters and correspondent units of international banks, what would give the interconnectedness of international financial centers. The banks are also an important measure for the definition of international financial centers because the presence of a headquarter—or subsidiary—increases both the number of competitors and the total size of the market, what makes these centers powerful places of attraction for an entire range of other correlated non-financial activities (Choi et al. 1986, 53).

The analysis of the financial assets from the 100 largest Latin American banks made it possible to establish a more detailed empiric assessment of the relative position of each country in the banking system of the region, in addition to a better approximation to understand the importance of Brazilian cities—and in special of São Paulo—in this banking hierarchy (Table 1.7).

The first characteristic that can be identified is the weight of Brazilian banks in the mass of assets under domain of the Latin American banking system, in the two dates shown by the table. When we consider the universe of 100 largest banks that the America Economy ranking provides, we observe that more than half of the Latin American banking assets is under the control of Brazilian banks. Mexico, the second largest economy in the region, represents around 15% of the total assets, with Chile in the third position of this hierarchy (with less than 10% in the 2 years contemplated). Even if added the assets of the eight largest Latin American economies (Mexico, Chile, Colombia, Argentina, Peru, Panama, Costa Rica, and Uruguay), it does not reach the amount equivalent to the total assets of Brazilian banks for the 2 years covered by the table.

Another important feature that the table shows is the impressive growth of banking assets in the continent: in the 16 years covered by the table, the total assets of the 100 largest banks in the region grew from around US$ 789,064,500.00 to US$ 3,095,845,500.00, nothing less than 292%. The most amazing growths were from two economies of smaller dimension in the region: Costa Rica (1,324%) and Dominican Republic (1,020%). Among the largest economies, it is highlighted the robust growth of assets of Colombian banks, which grew from only US$ 14,360,600.00 to US$ 154,304,600.00 (striking 974%), what made Colombia overtake Argentina in terms of banking assets, becoming the fourth most important financial economy in Latin America in 2016.

It is also verified that this prominence of Brazil increased in the last 15 years, once the participation of the country in the bank assets of the region grew 10% (from 50.4 to 60.4%); the reduction of the number of Brazilian banks among the 100 largest

Table 1.7 Latin America—Recent Evolution of Bank Assets—100 largest banks (2000–2016)

	2000			2016			Growth 2000/2016 (%)
	Banks	Total assets (US$ millions)	%	Banks	Total assets (US$ millions)	%	
Brazil	35	397,682.5	50.4	27	1,870,301.9	60.4	370
Mexico	11	123,705.8	15.6	15	403,541.6	13.0	226
Chile	12	66,398.8	8.5	10	292,554.1	9.4	340
Colombia	6	14,360.6	1.8	10	154,304.6	5.0	974
Argentina	19	133,956.2	17.0	9	98,233.7	3.2	−26
Peru	3	12,319.1	1.6	5	95,729.5	3.1	677
Panama	4	12,154.4	1.5	7	65,095.6	2.1	435
Costa Rica	1	2,017.1	0.3	4	28,724.7	0.9	1.324
Uruguay	2	8,691.0	1.1	3	25,346.4	0.8	191
Guatemala	0	–	–	3	23,313.0	0.7	–
Dominican Rep.	1	1,808.7	0.2	3	20,258.1	0.6	1.020
Ecuador	0	–	–	2	14,184.2	0.4	–
Paraguay	0	–	–	1	14.359,4	0.4	–
El Salvador	1	2,525.8	0.3	1	4,258.1	0.1	68
Venezuela	5	13,444.5	1.7	0	–	–	–
Total	100	789,064.5	100	100	3,095,845.5	100	292

Source America Economia (2000, 2016)

(from 35 to 27) shows also that continued to occur in the period an important process of banking concentration. Mexico had a slight fall in its participation, despite having increased the number of banks among the 100 largest (from 11 to 15 institutions). The negative highlight of the table is the performance of Argentina: from 2000 to 2016, there was a significant drop of its banking wealth, once the total assets controlled by Argentinian banks went from US$ 133,956,200.00 to US$ 98,233,700.00, a negative growth of −26%, removing the third place that the country had in the 2000 ranking (ahead of Chile and Colombia).

When we analyze closely the data about the distribution of headquarters of the largest Brazilian banks, it is possible to have a better idea of the importance of São Paulo as the main financial center in Brazilian territory (Table 1.8).

Table 1.8 Brazil—Main Commercial Banks and financial centers (2016)

LATAM ranking		Property	Name	City (headquarters)	Total assets (US$ mi)	Profits (US$ mi)
1°	1°	State-owned	Banco do Brasil	Brasília	451,146.0	2,977.2
2°	2°	Private national	Itau/Unibanco	São Paulo	402,901.4	5,683.4
3°	3°	State-owned	Caixa Econômica Federal	Brasília	369,430.6	1,212.2
4°	4°	Private national	Bradesco	Osasco (São Paulo Metropolitan Region)	290,896.7	5,029.4
5°	5°	Foreign	Santander Brasil	São Paulo	200,853.9	1,561.7
8°	6°	Private national	BTG Pactual	Rio de Janeiro	60,039.4	1,224.3
13°	7°	Foreign	HSBC	São Paulo	45,994.8	−449.5
17°	8°	Private national	Safra	São Paulo	41,354.4	505.3
20°	9°	Private national	Votorantim	São Paulo	32,892.3	118.3
24°	10°	Foreign	Citibank Brasil	São Paulo	24,709.6	220.1
27°	11°	State-owned	Banrisul	Porto Alegre	20,285.8	237.1
36°	12°	State-owned	Banco do Nordeste	Recife	12,987.7	137.3
38°	13°	Foreign	BNP Paribas	São Paulo	12,749.2	15.4
39°	14°	Foreign	JP Morgan	São Paulo	12,229.0	139.2
40°	15°	Cooperative	Sicredi	Porto Alegre	11,896.2	31.2
51°	16°	Foreign	Société Générale	São Paulo	9,381.4	−23.0
53°	17°	Cooperative	Bancoob	Brasília	9,229.4	43.3
57°	18°	Private national	Panamericano	São Paulo	8,343.5	−136.6
68°	19°	Private national	ABC Brasil	São Paulo	6,921.9	118.8
76°	20°	Private national	Daycoval	São Paulo	6,289.5	94.0
77°	21°	Foreign	Rabobank Brasil	São Paulo	6,280.9	56.9

(continued)

Table 1.8 (continued)

LATAM ranking		Property	Name	City (headquarters)	Total assets (US$ mi)	Profits (US$ mi)
78°	22°	Foreign	Deutsche Bank BR	São Paulo	6,262.7	−19.8
79°	23°	Foreign	BofA Merrill Linch	São Paulo	6,230.6	132.4
81°	24°	Foreign	Volkswagen	São Paulo	5,988.1	11.3
83°	25°	State-owned	Banestes	Vitória	5,633.4	51.5
87°	26°	Foreign	Tokyo Mitsubishi	São Paulo	5,245.6	32.3
91°	27°	Private national	BMG	São Paulo	4,981.8	20.2
Total					1,870,301.9	19,023.9

Source America Economia (2000, 2016)

As the table above shows, three of the five largest Brazilian banks (that are also the five largest banks in Latin America) have headquarters located in São Paulo, and control around US$ 894,652,000.00, i.e., 47.8% of total bank assets in the region. When we add all banks in São Paulo ranking among the 100 largest in Latin America (that are 19 in total), the amount of assets reaches US$ 929,644,000.00, what corresponds to 49.7% of totals. Another important data that the table shows is that all 11 foreign banks installed in Brazilian territory have their head offices located in the city of São Paulo (Santander, HSBC, Citibank, BNP Paribas, JP Morgan, Société Générale, Rabobank, Deutsche Bank, BofA Merrill Lynch, Volkswagen, and Tokyo Mitsubishi). This importance of the presence of foreign credit institutions in São Paulo (mainly banks) is also noted by statistics of Central Bank of Brazil. Since the year 2000, the percentage sited in São Paulo was always around 75% (Table 1.9).

These figures also identify that São Paulo has been consolidating itself as the main gateway city for international financial capitalism in the Brazilian territory, as already shown by Rossi and Taylor (2006, 2007).

Table 1.9 Brazil—Number of foreign credit institutions (2000–2016)

	2000	2016
São Paulo	55	61
Rio de Janeiro	7	5
Other cities	11	15
Brazil (total)	74	81

Source Banco Central do Brasil. IF.data—Data selected of supervised entities

São Paulo in the analysis of Global Financial Index (GFCI)

Following in this effort to define the general lines of the participation of São Paulo in the international financial division of labor, we also made an analysis of the data produced by the Global Financial Centers Index, created by the company Z/Yen Group Limited. The diagnostic proposed, in all editions, presents much more a preoccupation in the definition of the vision of global financial operators about the best places to make businesses, rather than a broader and stricter concern with academic or scientific purposes.[17] This basic diagnosis preoccupation explains, for example, why offshore financial centers appear better ranked than some important world cities (mainly cities from periphery, but also from the center of the world-system).

In the 2007' edition of the Global Financial Centers Index, the hierarchy of the main financial centers in the world was defined as follows: first: London (765 points), second: New York (760 points), third: Hong Kong (684 points), fourth: Singapore (660 points), fifth: Zurich (656 points), sixth: Frankfurt (647 points), seventh: Sydney (639 points), eighth: Chicago (636 points), ninth: Tokyo (632 points), and tenth: Geneva (628 points). The other centers would be ranked as follows (Table 1.10).

Due to the use of "competitiveness criteria" to define the most important countries in financial terms, offshore financial centers like Cayman Islands (16th), Hamilton (17th), and Channel Islands (19th) are found in a prominent position, ahead of other important cities. Even some "onshore" financial centers of smaller dimension obtain great prominence in the diagnosis, such as Sydney (7th), Vancouver (27th), Helsinki (32nd), Lisbon (42nd), and Athens (46th), cities that does not have this position in other kinds of analysis about "world cities" (that consider other more comprehensive factors to establish their rankings).

Another characteristic that draws attention int this ranking is the absence of financial centers from peripheral countries, in special the metropolises of major economies of Latin America (what interests us most in this book). The diagnosis of GFCI in the year 2008 shows for the first time the city of São Paulo in the 52nd position in the ranking and Johannesburg in 44th place. The cities of Rio de Janeiro and Buenos Aires are mentioned but are not part of the score due to not having enough amount of assessments to be ranked. The City of Mexico does not even appear in the ranking. In the Index of 2009, the ranking is enlarged for 75 cities, when also appears the concern of contemplating cities outside the world-system center: São Paulo would rank in the 41st position, Johannesburg—the only city in the African continent mentioned—the

[17] Although being an interesting work material, the research is clearly preoccupied in identifying the "comparative advantages" of London with the other cities in reference and once is sponsored by the proper City of London—and by the already mentioned Z/Yen Group Limited—has a clear bias and has to be read with great caution. The reports shall be considered only as an approximate measure of international financial centers, once it does not have as main concern to establish a critical and broad vision of the world cities from their financial "weight," but yet to define the hierarchy of "competitiveness" of the centers involved. It seems to reflect with fidelity only the vision of large "operators" of financial markets about the "advantages" that each financial center offers. Although the "competitiveness" criteria chosen by the Ranking are not precise enough to identify long-term processes—as Cassis (2018) argues—the results show interesting data on the hierarchy of the international division of labor in financial terms.

Table 1.10 World Financial Centres Rank Rating by GFCI (2007)

Rank	Name	Points	Rank	Name	Points
11th	Paris	625	29th	Stockholm	558
12th	Toronto	611	30th	Milan	546
13th	San Francisco	611	31st	Brussels	540
14th	Boston	609	32nd	Helsinki	537
15th	Edinburgh	605	33rd	Oslo	529
16th	Cayman Islands	604	34th	Copenhagen	525
17th	Hamilton (Bermuda)	603	35th	Vienna	518
18th	Melbourne	603	36th	Beijing	513
19th	Channel Islands	600	37th	Wellington	508
20th	Washington D.C.	594	38th	Rome	474
21st	Montreal	580	39th	Mumbai	460
22nd	Dublin	579	40th	Warsaw	460
23rd	Amsterdam	577	41st	Prague	453
24th	Shanghai	576	42nd	Lisbon	453
25th	Dubai	570	43rd	Seoul	434
26th	Luxembourg	570	44th	Budapest	425
27th	Vancouver	558	45th	Moscow	421
28th	Madrid	558	46th	Athens	395

Source Global Financial Centers Index no. 1 (2007, 13)

50th position, the City of Mexico the 55th, Rio de Janeiro 61st position, and Buenos Aires in 63rd place.

Since the 2010 edition, the Index makes an analysis about the macro-regions of the world, seeking a more detailed study from the following regionalization: 1. European centers, 2. Asian centers, 3. North American centers, 4. Middle Eastern centers, and 5. Offshore centers. In this year, therefore, was not included the consideration of the Latin American continent. This picture only changes with the incorporation, in the year 2012, of an assessment of "Americas," which brought the following ranking of Latin American financial centers in the list: São Paulo in 48th place, Rio de Janeiro in 52nd, Mexico City in 55th, and finally Buenos Aires in 68th position. As the rankings were made, they gradually increased the number of cities assessed: 80 in 2013, 83 in 2014, and 87 in 2016 reaching 96 in 2018. It was also maintained this minimum preoccupation in considering cities outside the central areas of the world-system. In case of Latin America, the positions in the different rankings varied as follows (Table 1.11).

Table 1.11 Latin American cities relative position as international financial centers (2012–2018)

	2012	2013	2014	2015	2016	2017	2018
São Paulo	48°	38°	38°	31°	51°	63°	67°
City of Mexico	55°	66°	70°	69°	73°	73°	70°
Rio de Janeiro	52°	31°	45°	35°	54°	82°	81°
Buenos Aires	68°	46°	25°	–	–	90°	75°
Panama	–	63°	59°	52°	63°	88°	80°

Source Global Financial Center Index (several years)

According to the analysis presented by Global Financial Index, therefore, São Paulo always configured as the main international financial center in the context of Latin America. For the Index, the dispute to attract international financial businesses is always between São Paulo, Rio de Janeiro, and Mexico City, with four offshore financial centers also with an essential role in the flow of resources in the region: Cayman Islands, Bermuda, Panamá, and British Virgin Islands (GFCI 2016, 2017, 2018).

References

Aalbers M (2017) Financial geography I: geographies of tax. Prog Hum Geogr: 1–12. https://doi.org/10.1177/0309132517731253

Amin S (1974). Accumulation on a world scale. A critique of the theory of underdevelopment. Monthly Review Press, New York

Anderson J (1991) A nova direita e a privatização: malogros britânicos, lições mundiais? Espaço & Debates 11(32):12–24

Arrighi G (1994) O Longo Século XX. Dinheiro, poder e as origens do nosso tempo. Contraponto/UNESP, Rio de Janeiro

Arroyo M (2006) A Vulnerabilidade dos territórios nacionais latino-americanos: o papel das finanças. In: de Lemos AG (ed) Questões Territoriais na América Latina. USP/CLACSO, São Paulo/Buenos Aires, pp 177–190

Ash S (2014) The development of submarine cables. In: Burnett DR, Beckman RC, Davenport T (eds) Submarine cables: the handbook of law and policy. Martinus Nijhoff Publishers, Leiden, pp 19–39

Bakis H (1984) Géographie des télécommunications. PUF, Paris

Bakis Henry (1987) Géopolitique de l´Information. PUF, Paris

Bassens D, Van Meeteren M (2015) World cities under conditions of financialized globalization: Towards an augmented world city hypothesis. Prog Hum Geogr 39(6):752–775

Bátiz-Lazo B, Wood D (2002) A historical appraisal of information technology in commercial banking. Electron Mark 12(3):1–12

Belluzzo LG (1995) O declínio de Bretton Woods e a emergência dos mercados globalizados. Economia & Sociedade 4(1):11–20

Beteille R (1991) La revolution boursière internationale. L'Information Géographique 55:1–10

BM&F Bovespa (2017) Annual report B3. https://ri.b3.com.br/enu/3756/RA2017-EN.pdf. Accessed 12 Aug 2018

Borchert JR (1978) Major control points in american economic geography. Ann Assoc Am Geogr 68(2):214–232
Braudel F (1958) Histoire et Sciences sociales: La longue durée. Annales. Economies, sociétés, civilisations 13(4):725–753
Budd L (1999) Globalization and the crisis of territorial embeddedness in international financial markets. In: Martin R (ed) Money and the space economy. Sons, New York, pp 115–138
Carvalho FCC, De Paula LF, Williams J (2015) Banking in Latin America. In: Berger A, Molyneux P, Wilson J (eds) The oxford handbook of banking, 2nd edn. Oxford University Press, New York, pp 984–1016
Cassis Y (2006) Capitals of capital. A history of international financial centres, 1780–2005. Cambridge University Press, New York
Cassis Y (2018) Introduction. A global overview from a historical perspective. In: Cassis Y, Wójcik D (eds) international financial centres after the global financial crises and brexit. Oxford University Press, Oxford, pp 1–15
Castells M (1996) The rise of the network society. Information age, vol I. Blackwell Publisher, Oxford
Chen K, Chen G (2015) The rise of international financial centers in mainland China. Cities 47:10–22
Chesnais F (1997) L'Émergence d'un régime d'accumulation mondiale à dominante financière. La Pensée no. 309:61–84
Choi SR, Tschoegl AE, Yu CM (1986) Banks and the world's major financial centers, 1970–1980. Weltwirtschaftliches Archiv 122(1):48–64
Choi SR, Park D, Tschoegl AE (1996) Banks and the world's major banking centers, 1990. Weltwirtschaftliches Archiv 132(4):774–793
Choi SR, Park D, Tschoegl AE (2003) Banks and the world's banking centers, 2000 (report). Rev World Econ 139(3):550–568
Clark G, Lai K, Wójcik D (2015) Editorial introduction to the special section: deconstructing offshore finance. Econ Geogr 91(3):237–249
Cohen RB (1981) The New International Division of Labor, Multinational Corporations and Urban Hierarchy. In: Dear M, Scott AJ (eds) Urbanization and urban planning in capitalist society. Methuen, London, pp 287–318
Cohen BJ (1998) The geography of money. Cornell University Press, Ithaca/London
Coob SC (1998) Global finance and the growth of offshore financial centers: the manx experience. Geoforum 29(1):7–21
Corbridge S, Thrift N (1994) Money, power and space: introduction and overview. In: Corbridge S, Martin R, Thrift N (eds) Money, power and space. Blackwell Publishers, Oxford, pp 1–25
Cordeiro HK (1986/87) Os principais pontos de controle da economia transacional no espaço brasileiro. Boletim de Geografia Teorética 16–17(31–34):153–196
Correa RL (1989) Concentração bancária e os centros de gestão do território. Revista Brasileira de Geografia 51(2):17–32
Crofts R, Siegler T (2018) Which tax havens are the most central? Applying social network analysis to understand firm service interaction. Fingeo. Financial geography working paper no. 20. http://www.fingeo.net/fingeo-working-paper-series/
Daniels P (1993) Services industries in the world economy. Blackwell Publishers, Oxford
de Brunhoff S (1967) La Monnaie Chez Marx. Éditions Sociales, Paris
de Brunhoff S (1996) A instabilidade financeira internacional. In: Chesnais F (ed) A Mundialização Financeira. Gênese, Custos e Riscos. Xamã, São Paulo, pp 35–59
Dias LC (1995) Réseaux d'information et réseau urbain au Brésil. L'Harmattan, Paris
Dicken P (2010) Mudança Global. Mapeando as novas fronteiras da economia mundial, 5ª edn. Editora Bookman, São Paulo
Dollfus O (1992) L'Espace financier et monétaire mundial. L'Espace Géographique 2:97–102
Dörry S, Robinson G, Derudder B (2018) There is no alternative. SWIFT as infrastructure intermediary in global financial markets. In: Fingeo. Financial geography working paper n. 22. http://www.fingeo.net/fingeo-working-paper-series/)

dos Santos T (1970) The structure of dependence. Am Econ Rev 60(2):231–236
Eichengreen B (1995) História e reforma do sistema monetário internacional. Economia & Sociedade 4(1):53–78
Eichengreen B, Lafarguette R, Mehl A (2016) Cables, sharks and servers: technology and the geography of the foreign exchange market. Eur Cent Bank Work Pap 1889:56p
Estado de São Paulo (2016) 11,5 milhões de documentos expõem corrupção global. https://politica.estadao.com.br/noticias/panama-papers,11-5-milhoes-de-registros-financeiros-expoem-corrupcao-global,10000024510. Accessed 31 May 2018
FSI. Financial Secrecy Index (2018) Tax justice network. https://www.financialsecrecyindex.com/. Accessed 14 June 2018
Fischer Gerald C (1968) American banking structure. Columbia University Press, New York
Freitas MCP (2011) A internacionalização do sistema bancário brasileiro. IPEA. Texto para Discussão 1566:84p
Freitas MCP, Prates D (2000) La experiencia de apertura financiera em Argentina, Brasil y México. Revista de la Cepal 70:53–69
Friedmann J (1986) The world city hypothesis. Dev Chang 17:69–83
Fröbel F, Heinrichs J, Kreye O (1980) The New international division of labour. Structural unemployment in industrialized countries and industrialization in developing countries. Cambridge University Press, Cambridge
Garbade KD, Silber W (1978) Technology, communication and the performance of financial markets: 1840–1975. J Financ XXXIII 3:819–832
GFCI. Global Financial Centres Index (Several Years). https://www.zyen.com/publications/public-reports/. Accessed 12 Jan 2019
Goldfinger C (1986) La Géofinance. Pour compreendre la Mutation Financière. Seuil, Paris
Gorostiaga X (1984) The role of the international financial centres in underdeveloped countries. Croom Helm, New York
Gottdiener M (1989) Crisis theory and socio-spatial restructuring: the US case. In: Gottdiener M, Komninos N (eds) Capitalist development and crisis theory: accumulation, regulation and spatial restructuring. Palgrave Macmillan, London
Gras NSB (1922) An introduction to economic history. Harper & Brothers, New York/London
Gras A (1993) Grandeur et Dépendance. Sociologie des macro-systèmes techniques. PUF, Paris
Guimarães SP (1999) Quinhentos anos de periferia. Uma contribuição ao estudo da política internacional. UFRGS/Contraponto, Porto Alegre/Rio de Janeiro
Guttmann R (1996) As mutações do capital financeiro. In: Chesnais F (ed) A Mundialização Financeira. Gênese, Custos e Riscos. Xamã, São Paulo, pp 61–96
Harvey D (1982) The limits to capital. Blackwell Publishers, Oxford
Hudson A (1999) Off-shores on-shore: new regulatory spaces and real historical places in the landscape of global money. In: Martin R (ed) Money and the space economy. Wiley and Sons, New York, pp 139–154
Hughes TP (1983) Networks of power. Electrification in western society, 1880–1930. John Hopkins University Press, Baltimore
Hymer S ([1972] 1983) Empresas Multinacionais: a internacionalização do capital. Edições Graal, Rio de Janeiro
ICIJ (2019) International Consortium of Investigative Journalists. https://www.icij.org/. Accessed 12 Feb 2019
Johnson HG (1976) Panamá as a regional financial center: a preliminar analysis of development contribution. Econ Dev Cult Chang 24(2):261–286
Joslin D (1963) A century of banking in Latin America. Oxford University Press, London
Kindleberger C (1974) The formation of financial centers: a study in comparative economic history. Princeton University Press, Princeton
Kirkman P (1987) Electronic funds transfer systems. The revolution in cashless banking & payment methods. Basil Blackwell, Oxford

Klagge B, Zademach HM (2018) International capital flows, stock markets, and uneven development: the case of Sub-Saharan Africa and the Sustainable Stock Exchanges Initiative (SSEI). Zeitschrift Für Wirtschafts-Geographie 62(2):92–107
Kurtzman J (1994) The death of money. How the electronic economy has destabilized the world's markets and created financial chaos. Simon & Schuster, New York
Labasse J (1974) L'Espace Financier. Analyse Géographique. Armand Colin, Paris
Laulajainen R (2002) Financial geography. A banker's view. Routledge, New York
Leyshon A (1995) Annihilating space? The speed-up of communications. In: Allen J, Hamnett C (eds) A shrinking world? Oxford University Press, Oxford, pp 11–54
Leyshon A, Thrift N (1997) Money/space. Geographies of monetary transformation. Routledge, London/New York
Lipietz A (1987) Mirages and miracles: the crisis of global Fordism. Verso, London
Machado LO (1996) O comércio ilícito de drogas e a geografia da integração financeira: uma simbiose? In: Castro IE et al (eds) Brasil: Questões atuais de reorganização do território. Bertrand Brasil, Rio de Janeiro
Machado LO (2017) O visível e o invisível: o sistema financeiro-corporativo mundial sob o prisma da extraterritorialidade e do binômio legal/ilegal. Geousp – Espaço e Tempo (Online) 21(2):325–340
Mackinder HJ ([1902] 1969) Britain and the British Seas. Greenwood Press, Westport
Malecki E, We H (2009) A wired world: the evolving geography of submarine cables and the shift to Asia. Ann Assoc Am Geogr 99(2):360–382
Mantoux P (1979) A Revolução Industrial no Século XVIII. Hucitec/UNESP, São Paulo
Marini RM (1976) Dialéctica da Dependência. Centelha, Coimbra
Martin R (1994) Stateless money, global financial integration and national economic autonomy: the end of geography? In: Corbridge S, Martin R, Thrift N (eds) Money, power and space. Blackwell Publishers, Oxford, pp 253–279
Martin R (1999) The new economic geography of money. In: Martin R (ed) Money and the space economy. Wiley, New York, pp 1–27
Mathelot P (1982) A Telemática. Edições, Lisboa, 70
Mattelart A (1992) Communication Monde. Histoire Des idées et des stratégies. La Découverte, Paris
Meyer DR (1986) The world system of cities: relations between international financial Metropolises and South American cities. Soc Forces 64(3):553–581
Nabarro W (2016) O mercado de capitais no território brasileiro: ascensão da BM&FBovespa e centralidade financeira de São Paulo (SP). Departamento de Geografia/Universidade de São Paulo, São Paulo (Unpublished master degree dissertation)
Nora S, Minc A (1978) L'Informatisation te la société. La Documentation Française, Paris
O'Brien R (1992) Global financial integration. The end of geography. The Royal Institute for International Affairs, New York
Palan J (2010) International financial centers: The British-Empire, city-states and commercially oriented politics. Theor Inq Law 11(1):149–176
Panreiter C (2015) Las ciudades latinoamericanas en la economía mundial: la geografía de centralidad económica y sus transformaciones recientes. Economía Unam 12(35):3–22
Peck J (2001) Neoliberalizing states: thin policies/hard outcomes. Prog Hum Geogr 25(3):445–455
Polanyi K (1944) The great transformation. The political and economic origins of our time. Farrar & Rinehart Inc, New York
Porteous D (1995) The geography of finance. Spatial dimensions of intermediary behaviour. Ashgate Publishing Limited, Aldershot
Porteous D (1999) The development of financial centres: location, information externalities and path dependence. In: Martin R (ed) Money and the space economy. Wiley, New York, pp 95–114
Reed HC (1981) The preeminence of international financial centers. Praeger Publishers, New York
Rixen T (2013) Why reregulation after the crisis is feeble: shadow banking, offshore financial centers, and jurisdictional competition. Regul Gov 7:435–459

Roberts S (1994) Ficticious capital, ficticious spaces: the geography of offshore financial flows. In: Corbridge S, Thrift N, Martin R (eds) Money, power and space. Basil Blackwell, Oxford, pp 91–115
Roberts R (2000) Por dentro das finanças internacionais. Guia prático dos mercados e instituições financeiras. Zahar Editores, Rio de Janeiro
Rossi E, Taylor P (2006) 'Gateway cities' in economic globalisation: how banks are using Brazilian cities. Tijdschrift voor Economische en Sociale Geografie 97(5):515–534
Rossi E, Taylor P (2007) Gateway cities: círculos bancarios, concentración y dispersión en el ambiente urbano brasileño. Revista Eure XXXIII nº 100:115–133
Sader E et al (eds) (2006) Latinoamericana - Enciclopédia Contemporânea da América Latina e do Caribe. Boitempo Editorial, São Paulo
Santos M (1975) Space and domination: a Marxist approach. Int Soc Sci J 27(2):346–363
Santos M ([1975] 1979) The Shared Space: the two circuits of the urban economy in underdeveloped countries. Methuen, London
Santos M (1994a) Técnica, Espaço, Tempo. Globalização e meio técnico-científico informacional. Hucitec, São Paulo
Santos M (1994b) Por uma Economia Política da Cidade. O caso de São Paulo. Hucitec, São Paulo
Santos M (1996) A Natureza do Espaço. Técnica e Tempo, Razão e Emoção. Hucitec: São Paulo. (Translated to French in La Nature de L'Espace, L'Harmattan, Paris, 1997)
Santos M (2000) Por uma outra globalização. Do Pensamento Único à Consciência Universal. Record: Rio de Janeiro. (Published by Springer in English, Toward an other globalization: from the single thought to universal conscience)
Santos M (2001) O País distorcido. Publifolha, São Paulo
Santos M, Silveira ML (2001) Brasil. Sociedade e território no início do século XXI. Record, Rio de Janeiro
Sassen S (1991) The global city. New York, London, Tokio. Princeton University Press, Princeton
Sassen S (1993) As Cidades na Economia Mundial. Nobel, São Paulo. (Published in English, Cities in a world economy)
Sassen S (1999) Global financial centers. Foreign Aff 78(1):75–87
Sassen S (2005) The embeddedness of electronic markets: the case of global capital markets. In: Knorr Cetina K, Preda A (eds) The sociology of financial markets. Oxford University Press, Oxford, pp 17–37
Scott S, Zachariadis M (2012) Origins and development of SWIFT, 1973–2009. Bus Hist 54(3):462–482
Simon C (1984) Les banques. La Decouverte, Paris
Sola Pool I (1993) Tecnología sin fronteras. De las telecomunicaciones en la época de la globalización. Fondo de Cultura Económica, Ciudad de México
Strange S (1994) From Bretton Woods to the casino economy. In: Corbridge S, Martin R, Thrift N (eds) Money, power and space. Blackwell Publishers, Oxford, pp 49–62
Strange S ([1986] 1997) Casino capitalism. Manchester University Press, Manchester
Sunkel O (1986) A Crise da dívida da América Latina. Dívida Externa e Empobrecimento. L&PM, São Paulo
SWIFT (2018) Society for worldwide interbank financial telecommunications. Annu Rev Powering Digit Econ: 21 p. https://www.swift.com/news-events/publications#topic-tabs-menu. Accessed 19 July 2019
Thrift N (1994) On the social and cultural determinants of international financial centres: the case of the City of London. In: Corbridge S, Martin R, Thrift N (eds) Money, power and space. Blackwell Publishers, Oxford, pp 327–355
Toussaint E (2002) A Bolsa ou a vida: a finança contra os povos. Perseu Abramo, São Paulo
UNCTAD (2017) Unctadsats. Country profiles. https://unctadstat.unctad.org/CountryProfile/GeneralProfile/en-GB/004/index.html. Accessed 02 Nov 2018
Vernay A (1970) Los Paraisos Fiscales. Plaza & Janes, Barcelona

Wainwright T (2017) Emerging onshore-offshore services: the case of asset-backed finance markets in Europe. In: Martin R, Pollard J (eds) Handbook on the geographies of money and finance. Edward Elgar, Cheltenham, pp 434–453
Wallerstein I (1976) Semi-peripheral countries and the contemporary world crisis. Theory Soc 3(4):461–483
Warf B (1989) Telecommunications and the globalization of financial services. Prof Geogr 41(3):257–271
Warf B (1999) The hypermobility of capital and the collapse of the Keynesian state. In: Martin R (ed) Money and the space economy. Wiley, New York, pp 227–239
Warf B (2002) Tailored for panama: offshore banking at the crossroads of the Americas. Geografiska Annaler Ser B Hum Geogr 84(1):33–47
Warf B (2006) International competition between satellite and fiber optic carriers: a geographic perspective. Prof Geogr 58(1):1–11
Warf B (2017a) Fiber optics: nervous system of the global economy. In: Warf B (ed) Handbook on geographies of technology. Edward Elgar, Cheltenham/Northampton, pp 113–125
Warf B (2017b) Digitalização, globalização e capital financeiro hipermóvel. Geousp – Espaço e Tempo (Online) 21(2):397–406
Werlen B (1993) Society, action and space. An alternative human geography. Routledge, London
Wójcik D (2011) The global stock market. Issuers, investors, and intermediaries in an uneven world. Oxford University Press, Oxford
Wójcik D (2012a) Where governance fails: advanced business services and the offshore world. Prog Hum Geogr 37(3):330–347
Wójcik D (2012b) The end of investment bank capitalism? An economic geography of financial jobs and power. Econ Geogr 88(4):345–368
Wójcik D, Knight E, Pazitka V (2018) What turns cities into international financial centres? Analysis of cross-border investment banking 2000–2014. J Econ Geogr 18(10):1–33
Zoromé A (2007) Concept of offshore financial centers: in search for an operational definition. IMF Work Pap 7, 1. https://www.imf.org/external/pubs/ft/wp/2007/wp0787.pdf. Accessed 09 Feb 2018

Chapter 2
Financialization of the Brazilian Territory

Abstract This chapter shows how the Brazilian financial and banking system has developed from a geographical perspective. The origin of the national banking and monetary system is discussed at a time when financial institutions were completely local or regional. Since World War II, but especially from the Reform of the National Financial System of 1964/65, a process of intensification of the financial division of labor began, and non-monetary institutions directly linked to the financialization of the territory (mainly investment banks and stock exchanges) arise and develop. The more regionalized character of the financial system is replaced by a national bank-based oligopolized structure, with its core area in the city of São Paulo. It is also analyzed the substitution of face-to-face/presential banking service channels to informational/remote ones.

Keywords Financialization · Financial and banking system · Regional banks · Bank topology · São Paulo · Market capital

2.1 When Finance Were Just Banking: Regional Banks and Currencies

The current chapter seeks to unveil the main characteristics of the evolution of the financialization of Brazilian territory, with emphasis in the analysis of its banking system, the true engine of national financial capitalism. Within the broader process of financialization, these two first items seek to identify more specifically the creditization of the territory and the logic of the diffusion of banking activity in the national space. Books and academic articles, in addition to official publications (especially those from Central Bank of Brazil), have not been concerned with highlighting what we may call the "geographic aspects" of these processes. Among these aspects that the current book intends to emphasize are as follows:

F. B. Contel, *The Financialization of the Brazilian Territory*, Economic Geography,
https://doi.org/10.1007/978-3-030-40293-8_2

1. The technical systems that are the basis of action of the banks:

As a great part of the contemporaneous literature about the geography of finance shows, the massive use of advanced technical systems—especially informational systems—has allowed the finances to become even more intense and omnipresent (Warf 1989; O'Brien 1992; Martin 1999). As also identified by Santos (1994, 1996), all institutions and firms, to act, depend on a set of technical objects and systems, and banks are not different. Just as financial agents needed submarine cables and artificial satellites to unify the world and make their action truly global (as we saw in Sect. 1.2), on a national scale, these technical networks were also necessary for scaling up their action.

2. The banking regions and their evolution:

It is also known, since the classic work of Jean Labasse (1955), that the development of banking activities is closely related to the regions that they act (Alessandrini and Zazzaro 1999; Alessandrini et al. 2009). According to their localization and complexity, commercial banks form their "banking regions" (Labasse 1955, 1974) or hinterlands (Porteous 1995, 1999). At each moment in history of countries, the nature and amplitude of these financial hinterlands modifies, and today depend heavily on information topology to form themselves. Historically, banking regionalism is one of the main expressions of the geographical organization of finance (Labasse 1974; Contel 2011; Guerim 2017), even if the phenomenon has recently lost expressiveness.

3. The role of the territorial normative contents:

A third central concern of this chapter is to emphasize the importance of the main legal norms that have conditioned the development of the Brazilian financial system. Laws, official rules, and standards of conduct can be understood as territorial normative contents, according to Santos (1996) and Mendes (2005).[1]

The literature about the genesis and development of banks shows that they appear in certain localizations for three main reasons: 1. existence in these areas of an expressive number of establishments and economic activities (agriculture, industries, or trade/service), 2. significant amount of economic agents (private firms, public institutions, or families/individuals) willing to consume banks' products and services, and 3. existence of "external economies" that are generated at the cities (Beaujeu-Garnier 1980, 238/239). Sheila Dow (1999, 32) shows also that the entire financial or banking system is a "product of its history," and based on her propositions, it is possible to define the evolution of the Brazilian financial systems as from three main periods, which would be (Table 2.1).

The "pure financial intermediation" phase in Brazil may be identified since the beginning of the banking system up to the first half of the past century. Throughout the

[1] According to Santos (1996), given the inseparability of forms and contents in the analysis of geographical space, norms can be understood as a constitutive content of space itself. Similar reasoning made earlier Anthony Giddens (1984)—not for the analysis of space, but society—from his theory of structuration, when he defines that norms and rules are an essential part of the structure of any society.

Table 2.1 Brazil—Periods and spatial implications of financial and banking development

Periods	Banks and geographic space	Events connected to the institutionalization of finances in Brazilian territory
First: From 1906 to 1964 Pure financial Intermediation	Serve the communities with wealth locally produced (wealth-based) and provide background for future financial centers	1906: Refoundation of Banco do Brasil 1921: Creation of General Inspectorate of Banks 1921: Installation of first public mortgage banks for agriculture financing 1945: Creation of Superintendence of Currency and Credit (SUMOC) 1952: Creation of National Development Bank (BNDE)
Second: **From 1964 to 1994** Creation of the Central Bank (lender of last resort) and capital market	Bank deposits are used as currency. Market depending of the trust dimension credited to banks. Banking systems develop a national level and start the creation of capital market	1964/65: Financial Reform and creation of laws that develop the capital market in the financial system 1964: Creation of Central Bank of Brazil and development of interbank loans 1976: Creation of the Securities Commission (CVM) 1986: Expansion of BNDES (as from BNDE) with voluptuous contribution of social security system
Third: **From 1994 to the present** Securitization	Deregulation opens for international competition, causing increase of concentration of activities in financial centers	1994: Implementation of Plano Real, privatization, and banking internationalization 2001: Installation of Brazilian Payment System (SPB) 2008: Merger of São Paulo Stock Exchange (Bovespa) with commodities and future exchange (BM&F)

Source Adapted from Dow (1999, 44) and Contel (2011)

entire nineteenth century, and up to the beginning of the twentieth century, several provincial and local banks were founded and extinguished, great part located in littoral cities, which were the main capital mobilization and accumulation centers (considering its participation in spatial circuits of commodities exportation such as sugar, cotton, and coffee, mainly). All these banks had a relative ephemeral life and closed their doors due to difficulties associated to low internal dynamism of economy (i.e., an economy dependent of external capitals), but also linked to the scarcity of currency in circulation, in which quantity was not enough for the dynamization of a more urban economy than was being created. It is emphasized that slavery was abolished in Brazil only on the year 1888, and it is from then on starts the formation of an actual labor market, with the diffusion of wages and the beginning of the modernization of labor regulation (Beiguelman 1977). Up to the end of the nineteenth century, with the proclamation of the Republic in 1889, there was not even one unified currency in the entire national territory; the Brazilian space was regionalized in six large monetary areas, with their respective issuing banks and regional currencies (Trigueiros 1966) (Map 2.1).

Apart from the creation of separate currency and banking regions, poorly integrated one another, the lack of physical integration of the territory would impede also the formation of a real national market (Barros de Castro 1972). The different Brazilian productive regions were "extroverted" (Kayser 1966), i.e., they were much more integrated to the commodities consumer centers in Europe than with themselves. The Saving Banks (*Caixas Econômicas*), another important financial institution at the time, followed likewise this eminently regional logic, as reminds Costa Neto

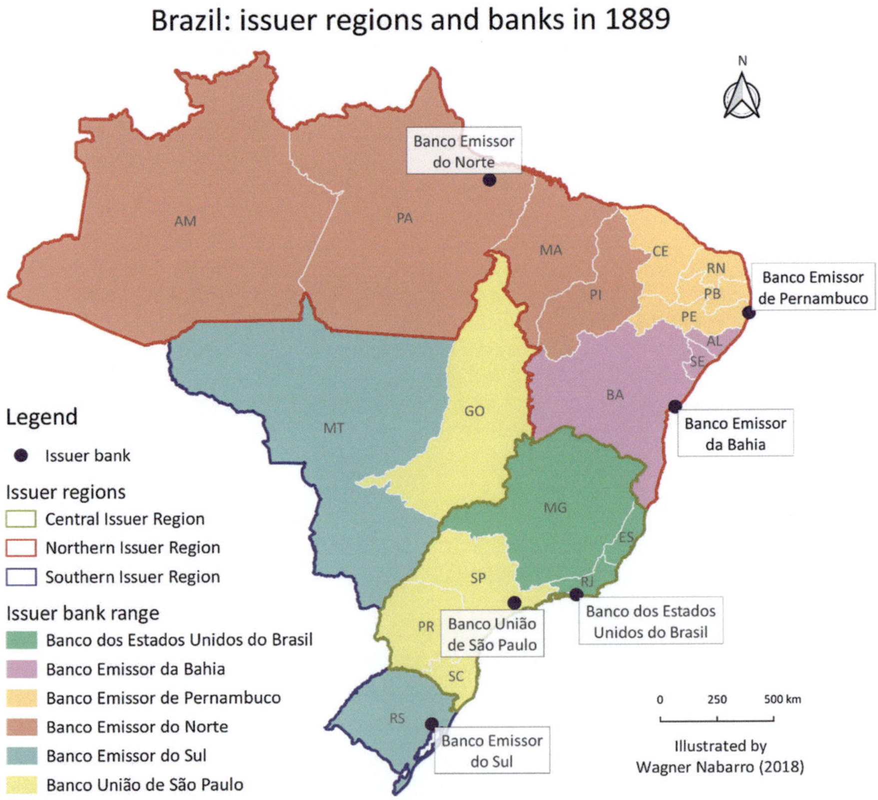

Map 2.1 Brazil—Issuer regions and banks (1889)

(2004, 14): "In each provincial state, the savings banks and charity funds (Montes de Socorro) had the same operational and management body, not having, however, any administrative or command interconnection between the institutions installed in the different provinces."

In the last decades of nineteenth century, there were only 68 bank branches in the country, and Rio de Janeiro concentrated around 80% of the deposits (Costa Neto 2004, 15). The banks were barely used by the economic agents to financial intermediation, i.e., for deposits, credits, custody, and valorization of economic resources.

In addition to these issuing and saving banks created with the purpose of serving as official entities for issuing currency—to help in the economic dynamism to all regions of the territory—a series of other banks appeared in a more "spontaneous" form, enjoying the most permissive moment of the monetary policy, which authorized to open bank branches without, however, these institutions having correspondent reserves in gold. This "anti-bullionist" (*papelistas*) policy—in opposition to the vision of "bullionists" (*metalistas*)—was, according to several studies in the period (Saes 1986), one of the main causes for the exaggerated banking expansion, which lead to the first severe financial crisis of the Republic, in 1896. As shown by Trigueiros (1966, 96), "this plethora of issuer banks, in the phase in which the republic regime started, generated one of the greatest financial crises which the Country had passed," the *Encilhamento*.

The Brazilian economy would only experience growth in the early twentieth century, when the territory begins to urbanize more intensely, and production circuits are formed more directly intended for national market. Some cities became surplus generation poles, capable of creating sufficient demand for the diffusion of banking phenomena to beyond coastal cities. As shown by Triner (1999, 133), "the monetized portion of the economy grew rapidly. The banking system increasingly accumulated and reallocated financial resources of the private sector at the expense of either personal or other institutional channels."

In this period, Rio de Janeiro was the main city in the country in economic and demographic terms, in addition to also being the political capital of the nation (since 1761). The banks created at this time were fundamentally commercial banks (also called "banking houses" or "casas bancárias") which served mainly "urban mercantile activities," in a local or utmost regional scale of operation. Considering the already mentioned "extroverted" operation of economy (based eminently on commodities exportation circuits), there was not enough impulse for diversification/complexification of financial activities. There was a less complex financial division of labor, and much part of the financing needs were supplied by non-financial agents: commissioners, import/export traders, and the producers themselves (self-financing) (Teixeira 2000, 26).

As from World War I, occurs an important process of "import substitution" that allows economy of the country to diversify and start a more dynamic industrialization process (Tavares 1964, 1978). The productive regions, still little articulated, become urbanized and turn the economy more complex. With regard to the banking system evolution, it may be said that on this period Brazil would enter in the "financial intermediation" stage, as preconizes Dow (1999). As highlighted in Table 2.1, the year 1906 may be considered a turning point in the history of the Brazilian financial structure especially by the changes that the State introduces in the banking system that was being formed. The main alteration was the "refoundation" of Banco do Brasil (for the third time), giving definitively to the bank the currency-issuing monopoly in national territory (Costa 2012). As a result of this change, the first public agencies of finance regulation were also created, once the number of banks started to grow again, the banking internationalization increased, what complexify the division of labor in the sector.

The first phase of the institutionalization of the regulation of the Brazilian financial system is established with the following events: 1. In 1921 was created the General Inspectorate of Banks (*Inspetoria Geral dos Bancos*, Article 3. No. 14.728 from 1921), which had since then the power to inspect the action of banks, with the "purpose of supervising the operational and patrimony conditions of banking institutions" (Chavantes 2007, 43). 2. In 1942 the Inspectorate would become the Banking Mobilization and Supervision House (*Caixa de Mobilização e Fiscalização Bancária*—CAMOB). 3. In the same year was created the Rediscounts Portfolio of Banco do Brazil (Decree 14.635, from January 21, 1921) or *Carteira de Redescontos do Banco do Brasil* (CARED), which would be a first attempt to create a lender of last resort in the banking system, as well as the authority that would require the banks to organize their social capital rationally, in addition to the minimum amounts deposited, in order to reduce the fragility of institutions, and of the banking system as a whole.[2] 4. In this same context was also created the Banking Clearing House that "would stimulate the use and acceptance of checks, increasing the liquidity by transactions speed increase in the settlement of debits and credits" (Chavantes 2007, 38).

The main consequences of this new banking regulation mechanisms and institutions were 1. to allow the expansion of use of scriptural money, 2. to reduce the need for banks to have their own deposits to lend money (which increased the possibility of credit supply), 3. to provide greater safety to the system, and 4. to strengthen national banks in case of crises/emergencies (which with rediscount have their vulnerability reduced, especially in respect to foreign banks, which were at that time more solid because of external funding) (Costa Neto 2004, 45). However, even with these alterations, the Brazilian banking system was constituted by a "large number of small and medium regional banks oriented for regional specificities, in a country of continental dimensions" (Cassiolato 1992, 173).

Concomitantly with these efforts aiming the modernization of Brazilian financial and banking system, a federal law, on the other hand, served as a powerful financial accumulation contention element. The Decree No. 22.626 from April 7, 1933, which regulated "*interest contracts and establishes other procedures*"—the so-called

[2]The importance of CARED is noticed by Eugenio Lagemann, when reminds that it was responsible for a "greater stabilization" of financing system as a whole, once "served as guarantee for banks, allowing them to put into circulation amounts that would otherwise be withheld" (Lagemann 1985, 35).

"Usury Law"—hindered financial operators to charge interests higher than 12% a year for operations contracted in national banking system. It ended up constituting a heavy discouragement for the development of the banking system and national market capital. It made difficult the formation of "voluntary savings" by the economic agents, mainly by the fact that the inflation rates during the 1940, 1950, and 1960 decades easily overcame 12% a year, causing to be negative the remuneration of most part of the financial assets available to the operators (especially savings accounts). Instead of investing their surplus and profits in financial assets, economic agents have channeled their resources to rural and urban real estate, jewelry, artworks, foreign currencies, among other types of non-financial assets (Costa Neto 2004).

In this period where contradictory measures were adopted by the federal government, it was only with the creation of Currency and Credit Superintendence (SUMOC—*Superintendência da Moeda e do Crédito*), by Decree No. 7.293, from February 2, 1945 that some functions of "monetary authority" were centralized in Brazil. SUMOC had as main concern the maintenance of the currency value, i.e., the inflation control, and its main attributions were 1. require issuance of paper money from Treasury Department, 2. define the percentiles to collect compulsory deposits from commercial banks, 3. set the rediscount rate (to interbank loans), 4. authorize the purchase and sale of gold and exchange linked securities, 5. instruct exchange policy, and 6. instruct supervision of banks (Galveas 1985, 29).

Even with the creation of SUMOC, significant part of the functions of monetary authority were divided with Banco do Brasil, which at that time was the largest Brazilian commercial bank (and State-owned). The Bank held the treasury of the Federal Government and the entire banking system. It also took care of the Check Clearance House and Banking Rediscount Portfolio, what made Banco do Brasil the main executor of "monetary authority" function (while SUMOC would be the main supervisor). The monetary authority functions, therefore, were divided between SUMOC and Banco do Brasil (Costa 2012).

Despite this institutional modernization which the financial system was being submitted at the time, it was not enough to definitively leverage the financing capitalism in Brazil. As shown by Maria da Conceição Tavares (1978, 129):

> The necessity of internal financial in primary-exporter model were basically associated to the commercial and urban development and were easily served by a banking network operating in little satisfaction conditions. The expansion of productive capacity in basic exportation activities or in infrastructure of public utilities services, although required a considerable accumulation of capital, did not pressure for long term credits to be supplied by a 'national' financial institutionalism.

This context of financial constraint has been modified since World War II. In the beginning of 1950 decade, Brazil changes definitively its development pattern, from an agriculture–export economy to an urban–industrial one. The industrialization process gains more dynamism and at the same time requires the creation of a set of larger public infrastructures—or "large technical systems" (Hughes 1987)—such as power generation plants (and major transmission systems), roadways, railroads, ports, airports, telecommunications infrastructures, among other complex and expensive engineering systems (*sistemas de engenharia*) (Santos 1996). It started to be selectively diffused in the territory a technical-scientific-informational *milieu*, i.e., this new built environment more intensive in science, technology, and invested capitals (which would replace a more "natural" geographic environment). This new technical basis of national territory is the main geographic cause of economic changes that would occur after World War II (jointly with the already mentioned movement of external investment expansion by multinationals, mentioned in Chap. 1). The greater technical densities of territory were a fundamental precondition to a heavier industrialization, an industrialization founded on large companies, being part of them multinationals. As analyzed by Peter Evans (1979) in his classic study, a "triple alliance" begun from this time, which reflects the division of labor that characterizes the industrialization process during practically the entire twentieth century:

1. *The Brazilian State* undertook the most onerous part of the division of labor, regarding production of capital goods (i.e., industrial inputs or the "Department I" of economy). It consists in companies like steel manufacturing plants, petrochemical industries, industrial inputs, and raw goods production, all demanding heavy transportation infrastructures (for the movement of industrial inputs and outputs) of telecommunications networks, as well as complex systems for availability of plentiful and cheap energy;
2. *The multinational companies* that were installed would take care of the production of durable goods, with greater added value, and more sophisticated technological level (products difficult to be produced by local/national companies because of their lower technological dynamism); and finally,
3. *The national firms*, which would take care of the production of simpler consumer goods, such as food, textile, and clothing products.

The urbanization also changes its features and accelerates. With the agriculture modernization brought by diffusion of this technical-scientific-informational *milieu* on the countryside, it was increased the number of people being expelled from their activities (once the modernization reduces the quantity of agricultural workers on field). The new generation that inhabits at the city needs, likewise, large infrastructures for their urban life, large technical systems associated with housing, water and sewer systems, electrification and public urban illumination, streets and sidewalks, schools, hospitals, among other elements of a built environment also more complex and diversified. It was these new territorial infrastructures that conducted the Brazilian financial system to modernize. The existing financial system until then was anachronic and inefficient to the new functions required. As shown by Teixeira (2000, 29):

> the changes in economy required new institutional mechanisms in the financial intermediation sector. The formation of financial resources for medium- and long-term credit became necessary, for public sector, and mainly for financing private companies, especially those that were already producing, in large scale, consumer durable goods

One of the main reflexes of these new—and more robust—financing necessities of the Brazilian economy was the creation of National Development Bank (Banco Nacional de Desenvolvimento Econômico—BNDE) in 1952. The Bank was founded by the president Getúlio Vargas (1882–1954), known for the developmentalist bias that characterized his government. The headquarter was installed in the city of Rio de Janeiro (where will remain until today), considering the fact that the capital of the Republic was also there (the capital would only be transferred to Brasília in 1956). On the year 1953, Vargas also created Petrobras, the State-owned company for the extraction, refinery, and commercialization of oil, which becomes a powerful economic development instrument in more national molds.

The country's political atmosphere was in this period strongly influenced by the ideology of "national-developmentalism" that saw the creation of a modern and robust industrial complex as one of the main paths for the dynamization of national economy, which would also make possible a less dependence of the Brazilian economy on the developed countries. It was necessary to undo the structural limitations (the so-called "bottlenecks"—*pontos de estrangulamento*) of the Brazilian national territory for the economic development to take off (Rangel 1980).[3]

The BNDE's action allowed the beginning of a long-term State economic planning, by instituting also the possibility of medium- and long-term financing, making it feasible to built infrastructures that needed more extensive financial amortization over time. Its creation and operation increase the centrality of Rio de Janeiro as the main national financial marketplace.

This developmentalist historical context was also important for allowing the creation of three federal banks specifically destined for financing economic activities "outside" the core area of Brazilian capitalism (especially outside the southeast region). These three banks would have as function to expand the financing possibilities of regions that were considered "less developed," especially in case of North and Northeast of Brazil. The federal banks created in this period were as follows:

1. *Banco do Nordeste do Brasil S/A* (BNB), created by Law no. 1949 from July 19, 1952;
2. *Banco de Crédito da Amazônia S/A*, created by Law no. 1184 from 1950. The Bank was a modernization of *Banco de Crédito da Borracha*, which had been created in July 1942 (to finance the production and exportation of rubber in Amazon), and has its denomination altered to *Banco da Amazônia S/A*—BASA— on the year 1966 (Law No. 5.122), name that carries until today;
3. *Banco Regional de Desenvolvimento do Extremo Sul* (BRDE), created in 1962, and that has as controllers the state governments of Rio Grande do Sul, Santa Catarina, and Paraná (Galvêas 1985, 53).

[3] As shown by the testimony of Cleantho de Paiva Leite (in Paiva 2012, 24), member of the BNDE's first Board of Directors, "The operating area of BNDES focused on the energy and transportation areas and, especially, basic industries. Energy to support large projects, such as the Hydroelectric Power Plants of São Francisco, Furnas, Cemig and others. Transportation, with emphasis on port and railroad structures and basic industries in the areas of steel manufacturing, heavy mechanics and chemistry."

Table 2.2 Brazil—Evolution of number of headquarters and branches of commercial banks (1940/1950)

	Headquarters	Branches	Branches/headquarters ratio
1941	512	1,134	2.2
1952	408	2,619	6.4

Source Correa (1989, 18)

Until this moment, therefore, the national financial system was constituted mainly by three types of agents: the SUMOC (created in 1945), the state and federal public banks, and private commercial banks. There was an explicit concern of federal governments that succeeded the conclusion of World War II in providing each region of the country with a proper banking system, in order to make available to local economic agents the financial products and services necessary for their own development.

In respect to the commercial banks network, between the decades of 1940 and 1950 there was an enormous banking expansion, due to three main factors: 1. a more subdue control by SUMOC, which imposed few financial and regulatory barriers to entry of new banks in the financial system; 2. the increase of demand for financial services, considering the urbanization and growth of the dynamism of the economy; and 3. the existence of elevated inflation rates that, combined with the Usury Law, obligated the banks to spread their network of agencies for frenetic capture of deposits (Passos 1973).[4] As shown by Correa (1989, 18), "The increase in the volumes of deposits remunerated with interest rates lower than inflation, resulted in the spatial expansion of the banking system, in an attempt to capture the maximum of deposits." In this context, the commercial banking system in Brazil presents the peak of its growth in a more decentralized feature—i.e., a system with a large number of smaller banks—in 1940 and 1950 decades.

As may be identified in Table 2.2, there is a double process of dispersion/concentration of banking system, once at the same time the number of banking headquarters was reduced (concentration process) and it was increased the network of bank branches (dispersion process). The commercial banks in this period served fundamentally for the deposits of surplus from urban population (individuals/households), for the provision of credits in shorter term, and most of all credit to meet agriculture financing and the necessities of private companies' working capital (industrial and commercial firms) (Galvêas 1985, 49). In this same year of 1941, the city of Rio de Janeiro, with 141 bank headquarters, was the main financial center in the country, also due to the fact of being the political capital of the nation. Because of the lower concentration of the sector and the great amount of smaller dimension

4 In order to reach the desired objective – to increase the volume of deposits received – the banks began the creation of new branches and elevated the number of employees. As could be imagined, this policy was, in certain way, one of the most important factors for elevation of their costs (Passos 1973, 56).

banks (with strong local and/or regional operation), this period may be defined as the *competitive period* of the Brazilian banking system. This more regionalized characteristic of the banking system would start to be transformed as from the Reform of National Financial System that starts on 1964 (as we will see in the item 2.2). The distribution map of commercial banks in Brazil in 1950 shows the primacy of Rio de Janeiro in the national financial system and the distribution of the bank branches network in the territory (Map 2.2).

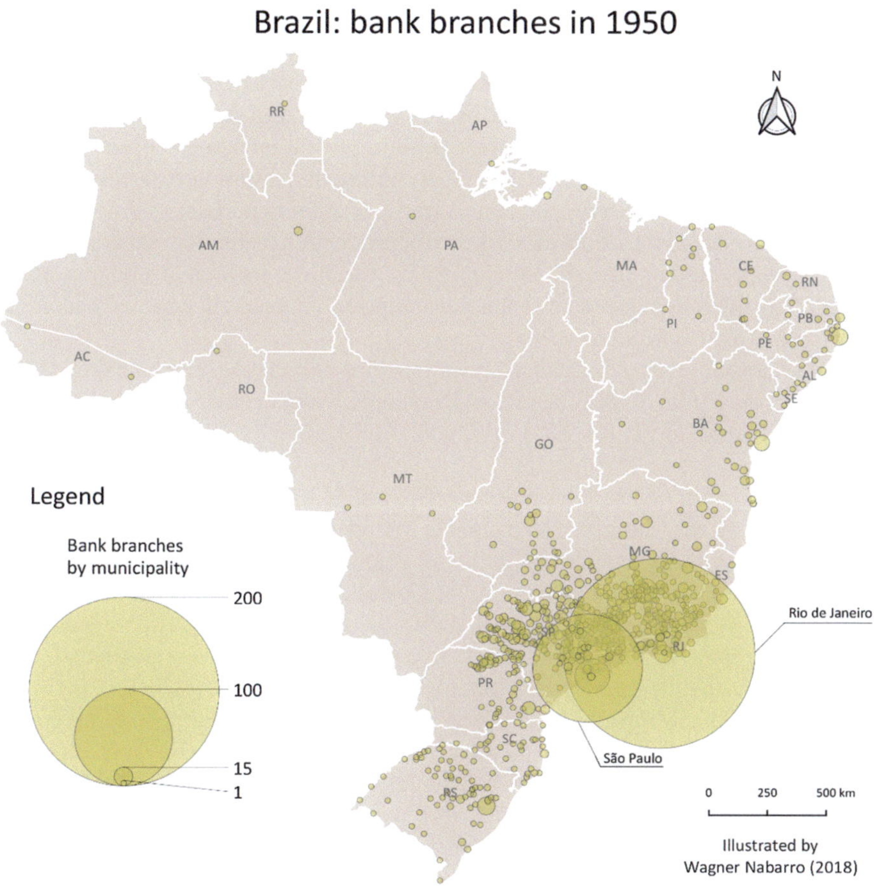

Map 2.2 Brazil—Commercial bank branches (1950)

2.2 The End of Regional Banks and the Primacy of São Paulo in the Urban Network

A vigorous transformation process of the national financial system took place in the years of 1964 and 1965. In these years—the first years of a military–dictatorial government—was implemented a deep change in the standards, laws, and in the institutionalization of the national financial system, through the conventionally called Reform of National Financial System (Reforma do Sistema Financeiro Nacional, RSFN). This Reform was concretized by the enactment of two main Laws: Law No. 4.595 from January 31, 1964 (which became known as *Banking Reform Law*) and Law No. 4.728 from July 14, 1965 (the so-called *Law of the Capital Market*).

The processes initiated after World War II deepened. The national industrial complex becomes more robust and dynamic, with important industries in practically all segments necessary for the national economy autonomy/sovereignty (steel manufacturing, petrochemical, heavy mechanics, transportation materials, industrial equipment, etc.). Added to these industries those for production of durable and non-durable goods, such as automobiles, home appliances, furniture, food, and products which had an expanded demand due to the consumption increase of families becoming urbanized (Castro and Souza 1985).

The urbanization was the main force of territory remodeling, and gains renovated impulse. From a total of only 36.16% of the people living in cities in the 1950 decade (within a universe of 51,944,397 inhabitants), this percentage changes to 44.67% in 1960 (from a universe of 70,992,343 inhabitants) reaching at 1970 with 55.92% of Brazilians living in cities (in a universe of 94,508,583 people) (IBGE 2018).

The territory was submitted to complete remodeling of its transportation system, in great part provided by the substitution of a model founded on railway systems for roadway movement systems (Barat 1978; Contel 2001). It is the broad diffusion of a network of modern paved roads that allows the formation of an effective national market, with more integration of productive regions that were less connected with themselves during the last decades (Santos e Silveira 2001).

In addition to new infrastructures for physical transportation of people and goods, there was also installed in the territory new technical systems destined for information transmission, networks that were fundamental to the operation of financial firms. It is as from the Military Coup of 1964 that were installed large technical systems linked to the circulation of information in the territory, such as telephony, radio transmission, and television broadcast networks, and most of all the diffusion of information transmission forms by artificial satellite, technology available from 1969 in Brazil (Dias 1996). This new material basis allowed the development of what may be called "remote action" (*tele-ação*) (Santos 1996) in the largest Brazilian companies, i.e., it was not necessary to the firms to be physically co-present with each other, making possible that the decision points of the economy (headquarters) were away from the *executors of orders* (the units of production), and by this way the "material separation of production activities and organizational unification of commands" was allowed (Santos and Silveira 2001, 73). This remote action, on its turn, is what permitted the banking concentration process to take place in Brazil, once through them large national banks could be formed, and these larger banks could then buy smaller (local and regional) banks, in different parts of national territory (as mentioned).

Under the point of view of its internal operation, the banking firms started an automation process in this period, still incipient. Until then, the machines and devices used by the banks had low information processing capacity, were highly dependent of workforce (i.e., demanded great amount of bank employees for operation), and they were mechanically driven. Objects such as tabs, typists, panels and blackboards, and punch cards were part of this predigital information environment (FGV 2010, 44).

According to Hindenburgo Pires (1997), among the main changes occurred in 1960 decade, one was the introduction to new technical systems that allowed the centralization of information processing, mainly the creation of Data Processing Centers (DPC), the introduction of new hardware devices (minicomputers) and large telecommunication systems. These technical systems, on their turn, allowed the passage of a *disperse* form for a *centralized* form of financial institutions information processing (Pires 1997, 4).

The first computers used in the Brazilian banking system were installed early in the 60s and had as characteristic a very limited processing capacity (when compared to the current systems) and were used mainly to "accelerate the processing of a growing volume of information produced in the branches" (FGV 2010, 61). More precisely, it was Banco Bradesco that in 1962 purchased and installed their first computer, used for the automation of back-office operations.

Concomitantly to those changes of technical characteristic, fundamental transformations occurred also in the normative aspects of banking regulation. As mentioned, the Reform of the National Financial System aimed the modernization of the financial institutions and regulation, mostly due to the increase of "specialization of financial institutions"; a compartmentalized model was adopted between different types of operators, with clear delegation of functions to *regulators* (the public monetary authorities) and *operators*, publics, and private (commercial banks, investment banks, stock exchanges, insurers, credit cooperatives, etc.). This compartmentalization also reflected on the division of labor between financial operators: contrary to a model based on "universal banks," rules were established for each type of firm to have a specific market niche, model inspired in great part on US legislation of Glass–Steagall Act from 1933 (Carvalho et al. 2015).

Within this spirit, the main modifications introduced in the national financial system were as follows:

1. Creation of the Central Bank of Brazil (BCB) that would constitute the maximum monetary authority in the country (in substitution to the SUMOC);
2. Creation of a set of institutions and standards aiming to modernize the national capital market, starting with the creation of National Monetary Council (1964) and afterward by the proper regulator agency of capital market, the Securities and Exchange Commission (*Comissão de Valores Mobiliários*, CVM, created in

1976), which would be the federal agency responsible for regulating the action of capital market operators, such as stock exchanges, investment banks, finance companies ("financeiras"), brokers, etc. (entity equivalent to US Securities and Exchange Commission—SEC);
3. Institution of the *monetary correction* ("correção monetária"), an accounting indexation mechanism—or "automatic" updating—of financial values transacted by means of formal contracts, that aimed to impede contracted values to be depreciated due to the elevated inflation rates (incorporating the inflation variation in the contracts);
4. Creation of the Habitation Financing System (*Sistema Financeiro da Habitação*, SFH) and National Bank of Habitation (*Banco Nacional da Habitação*, BNH), both focused on one of the main problems caused by urbanization: housing deficit among the lowest income population.

Another important inflection point in the Brazilian financial system occurs in the year 1988, in the context of promulgation of new Federal Constitution. By the new laws consubstantiated by the *Magna Carta*—and especially by the modifications of the financial regulation proposed by the Central Bank—occurs what was called the "mini-financial reform," which had a fundamental aspect: the legal authorization for commercial banks to act on other portfolios, previously reserved to other types of institutions. From the edition of the Resolution 1524, dated September 21, 1988, there were some modifications in the financial system that until then was more compartmentalized and formed by specialized institutions (conditioned by regulation to practice a restricted range of operations). With this Resolution, it was given the authorization for large commercial banks to officially become "universal institutions."

With the implementation of these "mini-reform" alterations in 1988, the first consequence was the expressive growth of the number of banks in operation. There was an increase from 104 banks in 1988 to 244 at the end of 1994. In this occasion, some brokers and other financial operators transformed into commercial banks, increasing the number of market agents.

At this time, the other central transformation of Brazilian financial system was the promulgation of Real Plan (Plano Real), in 1994. Real Plan was a monetary stabilization program that aimed fundamentally to control inflation levels of the national economy. Under the macroeconomic point of view, the measures had a clear neoliberal basis, once the inflation control was followed by a rigid foreign exchange policy (to maintain an external fix value for the currency) and of adjustment measures such as to open the economy for foreign investments, privatizations, and fiscal adjustments (following the Washington Consensus "commandments" mentioned on item 1.1).

In respect to banking sector, the new Plan—and the set of policies that were implemented to support it—had direct consequences that lead to a greater consolidation of banking system. The centralization of banking capitals occurred by two main processes: privatization of several important public banks (especially public banks at the state level), and purchase of small- and medium-size private banks by larger institutions (with their national dimension already consolidated). This banking consolidation meant also, as we have seen, a reduction of banking regionalism in Brazil that is still going on.

From the point of view of internal operation of banks, the first change resulting from Real Plan was precisely the loss of the main banking income source until then, which were called the "inflationary profits" (Alexandre et al. 2006). In 1993, the total inflation of the economy was around 2477%/year; in the month exactly before starting the Plan (June 1994), the inflation dropped to 47.43%/year. With high levels of inflation, the banks obtained very high profits through this main mechanism: the investment of their liabilities (deposits) in operations and papers tied to high inflation rates (which several federal bonds offered) combined with a low remuneration of these liabilities (mechanism also known as "floating").

In respect to the commercial banking system, the inflation control changed completely the logic of their operations. The "floating" profits ended up hiding the operational deficiencies that banking institutions presented (mainly internal operating costs). In this context, banks started to worry more with the efficiency of their internal structures and seek for more economies of scale, once the operation costs become more easily identified, and charging fees from their clients became one of the fundamental sources of banking revenue. It should be emphasized that this search for an increase of efficiency meant also two things, in the period after the implementation of Real Plan: the greatest automation of the banking system as a whole and the dismissal of great amounts of bank workers (FGV 2010, 228).

Other fundamental modifications occurred with the publication, by the Central Bank, of the Exposure of Reasons 311 (Exposição de Motivos 311), from August 23, 1995 (signed by the Treasury Minister Pedro Malan). According to the document, the entrance of foreign banks in the national market would "open" the Brazilian economy, capture savings from abroad and "would introduce new technologies" in the sector, advocating therefore about the necessities of "increase the percentile of participation of individuals and legal entities, resident or domiciled abroad, in the capital of national financial institutions" (BCB 1995). Previously, the foreign banks that operated in the country used the reciprocity rule, i.e., the national monetary authorities allowed for the foreign banks to install here if they also allowed the installation of Brazilian bank branches in their territories. The control/purchase of national institutions by non-residents, until the publication of Exposure of Reasons 311, was forbidden.

In addition to the Exposure of Reasons No. 311, the Stimulation Program to Restructuring and Strengthening the National Financial System (PROER—*Programa de Estímulo à Reestruturação e ao Fortalecimento do Sistema Financeiro Nacional*) from November 1995 and the Incentive Program to Reduction of State Public Sector in Banking Activity (PROES—*Programa de Incentivo à Redução do Setor Público Estadual na Atividade Bancária*) from August 1996 were the other two public policies implemented that modified substantially the ownership structure of the national banking system.

The Stimulation Program for Restructuring and Strengthening the National Financial System (PROER), created in 1995 through Resolution No. 2208 from 11.03.1995 from the National Monetary Council (CMN), was an efficient stimulation and financing mechanism for banking concentration in Brazil. It fundamentally aimed to reorganize and sell private banks that had become more fragile under the new macroeconomic conditions (mainly the absence of high inflation rates). According to Fernando Barbosa (2007, 103), PROER "allowed Central Bank of Brazil to grant financing for reorganization of financial institutions and increased the powers of Central Bank to transfer rights and obligations in the intervention processes, through corporate reorganizations, by secession, merger, or incorporation processes." The studies that analyzed this banking restructuring promoted by the Federal Government at the time did not explain one of their main characteristics: the loss of regions over the control of their own commercial banks, and the increase of primacy of Metropolitan Region of São Paulo in the control of the national banking system.

As shown by the work of Matias (1999), in the period from 1994 to 1998, the activities of 11 large national private banks were ended, 4 of them had headquarters outside São Paulo. The four main private banks of national dimension that were not headquartered in São Paulo were Banco Nacional (with headquarter in Belo Horizonte, Minas Gerais), Bamerindus (with headquarter in Curitiba, Paraná), Banco Econômico (with headquarter in Salvador, Bahia), and Banco Banorte, with headquarter in Recife, Pernambuco.

Once mentioned the main consequences of PROER for banking restructuring, it is necessary to highlight the consequences of the other Plan, the Incentive Program for Reduction of State Public Sector in Banking Activity (PROES), implemented through Provisional Decree No. 1.514, from August 7, 1996. For Vidotto (2005, 75), PROES has as "main purpose" the privatization, extinction, or transformation of state-owned banks into development agencies, "providing for the provision of resources for the states to refinance its debts with their banks." PROER was one of the main neoliberal measures to reorganize national public accounts, by allowing an expressive privatization of public banks at the state level of the federation.

With PROES, the following commercial and saving banks were liquidated: *Banco do Estado do Acre*—Banacre, *Banco do Estado do Amapá*—Banap, *Banco do Estado do Rio Grande do Norte*—Bandern, *Banco de Desenvolvimento do Rio Grande do Norte*—BDRN, *Banco do Estado do Mato Grosso*—Bemat, *Banco do Estado de Rondônia*—Beron, *Caixa Econômica do Estado de Goiás*—Caixego, *Caixa Econômica do Estado de Minas Gerais*—Minascaixa, *Banco do Estado de Alagoas*—Produban, and finally *Banco do Estado de Roraima*—Baner (Salviano Jr. 2004; Contel 2011). Except for *Minas Caixa*—located in the state of Minas Gerais, the third wealthiest state of the federation—all other banks liquidated were located in more peripheral states (mainly in the North and Northeast regions), distanced from the core area of national capitalism. This process is a strong element of the loss of financial dynamism of the most peripheral regions of the territory.

In addition to the banks that were simply extinct, another significant number of them were bought by large national or international banks. The list of state-owned banks bought in the PROES ambit is provided in Table 2.3.

With the closing and/or privatization of these banks, the more peripheric areas that had their own financial institutions lost their capacity to control finances regionally.

Table 2.3 Brazil—Privatized/purchased government-owned banks (1998–2006)

Name of public bank purchased	No. of branches	Buying bank	New head office localization
Banco do Estado de São Paulo—Banespa	1,330	Santander	São Paulo
Banco do Estado de Minas Gerais—Bemge	455	Itaú	São Paulo
Banco do Estado do Paraná—Banestado	376	Itaú	São Paulo
Banco do Estado do Rio de Janeiro—Banerj	190	Itaú	São Paulo
Banco do Estado da Bahia—Baneb	170	Bradesco	São Paulo
Banco do Estado de Goiás—BEG	151	Itaú	São Paulo
Banco de Crédito Real de Minas Gerais—Credireal	86	Bradesco	São Paulo
Banco do Estado do Maranhão—BEM	76	Bradesco	São Paulo
Banco do Estado do Ceará—BEC	69	Bradesco	São Paulo
Banco do Estado de Pernambuco—Bandepe	52	ABN/Amro	São Paulo
Banco do Estado do Amazonas	36	Bradesco	São Paulo
Banco do Estado da Paraíba—Paraiban	16	ABN/Amro	São Paulo

Source Salviano Jr. (2004), Contel (2011)

After the Real Plan, therefore, the more regional characteristics of the Brazilian banking system were almost definitely terminated. The concentration of bank control was followed by an expansion of the network of bank branches. With the number of banks reduced, the power of banks located in São Paulo became even bigger. By adding the privatizing character of measures adopted in the context of Real Plan, with its liberalization to purchase local banks by foreign ones, the result could not have been different: the Brazilian banking system becomes more concentrated and internationalized (Table 2.4).

Table 2.4 Brazil—Evolution of banking control after *Plano Real*—in % (1988–2000)

	1988	2000
Banks with foreign control	9.62	33.11
Private national banks	56.85	42.56
Public national banks	33.53	24.33
Total	100	100

Source Central Bank of Brazil 2018 (Available at http://www.bcb.gov.br/htms/deorf/e88-2000/texto.asp?idpai=RELSFN19882000)

2.3 The Core of Financialization: Capital Market, Investment Banks, and Stock Exchanges

As mentioned, Brazilian financial growth intensified after the end of World War II, due to the robust industrialization and urbanization of the territory. In addition to the process of bank concentration under way and the creation of the BNDE in 1952, other institutional measures were taken by successive Federal Governments to create a more dynamic capital market in the territory. This capital market allowed the channeling of economic surpluses (savings) to productive investments, a basic rule of economic development in any country (Rangel [1957] 1990; Furtado [1967] 1987).

One of the first steps in this direction was taken in 1946, when the first Credit, Financing, and Investment Societies (popularly known as "*financeiras*") were allowed to operate in Brazil. They can be considered the first non-monetary financial institutions in the territory and were created "due to the impossibility of the banking system offering capital for the medium and long term" (Passos 1973, p. 34). These companies started to issue financial bonds to the public and were initially divided into two main types:

1. *Credit and Financing Companies*, aimed at offering credit for consumption (especially durable goods), as well as working capital for companies;
2. *Investment Companies*, which began to create and manage investment funds, and to apply these funds to securities on Stock Exchanges (Passos 1973, pp. 36/37).

The initial importance of "*financeiras*" was also highlighted by the fact that, through a basic accounting procedure (the contracts substituted the denomination "interest" for "discount in relation to the par value of the security"), these companies could partially circumvent the Usury Law, remunerating assets placed on the market with real positive values (Teixeira 2000).

Throughout the 1950s, the Brazilian financial system was still be relatively simple, with no actors—neither private nor public—capable of supplying the productive sector with the longest term financial resources with adequate interest conditions to spur economic growth. The financial system was heavily based on the network of existing commercial banks, which had neither the financial capacity nor the political interest to offer products and services in accordance with the new financing needs brought about by the urbanization and by the vigorous and ongoing industrialization of the territory.

The stock market "practically did not exist" in this period (Vital 1976, 130), mainly due to the tradition of large Brazilian family-owned and privately traded companies, which prevent the launching of primary bonds in the form of shares. Unlike the United States, where industrialization soon enjoyed a powerful relation with the capital market to finance the expansion of productive capacity, in Brazil, less complex, short-term mechanisms of self-financing remained. The true capitalization of firms thus depended on their "capacity for self-accumulation of surpluses" (Vital 1976, 130), which took the dynamism out of growth.[5]

A brief description of the state of the capital market in Brazil in the 1950s could be this:

1. There were virtually no "non-monetary indirect bonds" on the financial market;
2. The "*Financeiras*" were the only existing non-monetary private institutions;
3. The public institutions—the National Economic Development Bank (BNDE) and the National Cooperative Credit Bank (BNCC)—were the only non-monetary institutions to promote the country's economic development;
4. Stock exchanges (i.e., the primary market) "showed poor performance as a source of funds" (Vital 1976, 137).

Table 2.5 provides a more approximate measure of how the national financial system evolved in terms of two main indicators: the ratio of financial assets/gross domestic product and the ratio of non-monetary assets to financial assets.

A first information that can be highlighted in the Table 2.5 is that, between the years 1947 and 1966, there was a fall in the proportion of financial assets/GDP, which can be credited to all these difficulties already mentioned (and hindered the financialization of the Brazilian economy). Non-monetary assets have similar behavior. Given the aforementioned negative effect of the Usury Law on the economy of the period, the preference of the surplus spending units—the investors—shifted from financial assets to several other types of tangible assets, such as real estate, jewelry, works of art, and also foreign currencies (as already mentioned) (Vital 1976, 136).

The new institutional framework brought with the Reform of the National Financial System was also fundamental for expanding the mechanisms for draining resources from the economy to compose internal public savings, aimed to reinforce the necessary investments in the productive system. The main institutions created for capital market development included the following:

1. National Bank for Economic Development (BNDE—*Banco Nacional de Desenvolvimento Econômico*) (In addition to the BNDE, 9 other provincial development banks are created between 1965 and 1972);
2. Investment Banks;

---

[5]Self-financing, in turn, consisted of three main techniques: 1. Foreign exchange financing (devaluation of the exchange rate and easy importation of production assets); 2. inflation financing (use of price differences in economic transactions in favor of the company); 3. promissory notes to manage payment deadlines for suppliers.

Table 2.5 Brazil—Monetary and non-monetary assets in the evolution of the financial system (1947–1974)

Year	Financial assets/GDP (%)	Non-monetary assets/financial assets	Year	Financial assets/GDP (%)	Non-monetary assets/financial assets
1947	40.5	30.5	1961	30.5	10.1
1948	36.4	29.8	1962	29.4	7.9
1949	35.3	28.1	1963	26.3	7.2
1950	37.8	23.7	1964	25.0	7.6
1951	40.5	20.4	1965	29.4	13.2
1952	36.7	18.3	1966	24.9	20.2
1953	36.1	16.5	1967	29.5	27.0
1954	32.3	15.1	1968	32.2	32.6
1955	29.6	13.0	1969	33.7	36.2
1956	27.7	11.5	1970	37.0	43.2
1957	29.5	9.8	1971	39.8	50.0
1958	29.5	8.7	1972	49.7	55.0
1959	29.5	6.9	1973	53.0	56.4
1960	29.4	7.7	1974	48.9	56.6

Source Vital (1976, 135)

3. Financing Companies (the new "*financeiras*");
4. Real Estate Credit Companies;
5. Savings and Loan Associations;
6. Brokerage Firms and Securities Dealers (Passos 1973).

In this context, the National Housing Bank (BNH—*Banco Nacional da Habitação*) was created by the Law No. 4,380 of August 21, 1964, and in a few years "grew in importance within the Brazilian capital market" (Passos 1973, 80). The main function of BNH, which operated from 1964 until its elimination in 1986, was to guide and regulate the Housing Finance System (SFH—*Sistema Financeiro da Habitação*). Given the huge volumes of resources required for investment in housing and urban infrastructure, the National Bank for Housing played a key role in this financing. The two main programs—or set of "problems"—that the bank implemented, were as follows:

1. Urban development programs, aimed at the implementation and improvement of a series of public infrastructures in cities, such as water and sewage systems, public lighting and energy equipment, etc.;
2. Housing programs related to the construction and commercialization of housing units (for middle- and lower income population) (Oliveira 1979, 68).

The creation of the National Housing Bank and all other non-monetary institutions mentioned above help to explain the dynamism that the capital market gained after the

Reform. As Vital (1976, 141) stated, in this period "the capital market, which had been dwindling over the years, and concentrating on short-term operations, grew in size, innovated with the emergence of innumerable non-monetary financial institutions and began to offer to the borrowers and savers new and complete menu of financial opportunities." This dynamization can be empirically verified by analyzing the non-monetary financial inventories in relation to the Gross Domestic Product (GDP). This percentage increased from 3.2% in 1965 (year that the financial system reform was implemented) to 8.6% in 1968, reaching 13.1% in 1971, 23.1% in 1974, and 30.8% in 1980 (Castro 1979, p. 40). This growth is also explained by the institutional incentives that the Federal Government implemented at the time to stimulate the development of the Brazilian capital market. One of the main incentives was Decree-Law no. 157 of 1967, which created mutual investment funds (the so-called "Funds 157"), that allowed the public to apply 10% of their income tax due on the purchase of securities from these funds. These papers were intended both to "enable the capitalization of companies through the stock market" and "to induce members of the community, both individuals and corporations, to accept the stock market as a safe and competitive alternative to apply their savings" (Waddington 1976, 186).

In summary, the main changes in the post-1964 capital market in Brazil included as follows:

1. Creation of several non-monetary financial intermediaries with well-defined roles and functions (given the principle of compartmentalization that guided financial reform);
2. Rapid growth of non-monetary assets in relation to the country's Gross Domestic Product;
3. Change in the behavior of economic agents, especially the final borrowers, who chose to prefer indirect (non-monetary) bonds and securities rather than direct bonds (also called "primary" bonds);
4. Increase in the average term of the loans, with the financial innovations (new equities) introduced in the economy;
5. Expansion of the transactions on the Stock Exchanges (mainly in Rio de Janeiro and São Paulo);
6. Formation of large financial conglomerates, led by commercial banks that acquired greater size and financial strength during the period (some having reached the size of national banks) (Vital 1976, 141/142).

Hence, the 1964/65 Reform allowed the beginning of the securitization of the economy, since it increased the share of national wealth represented by bonds, stocks, certificates of deposit, among other securities that have liquidity and profitability since the 1964 changes. In the words of Tavares and Carvalheiro (1985, 35), there was a very rapid transition in the nature of the composition of "financial assets," which went from "monetary" to "non-monetary" ones:

> Monetary assets accounted for 95% of the financial assets of the economy in 1961/63, thus confirming that the financial system in Brazil until that year was mostly restricted to the banking sector. From 1964/65, with the reforms introduced in the financial system, the monetary segment continually lost its share, falling to the level of 24%, in 1981, of total

Table 2.6 Brazil—Volume of stocks traded on the Rio de Janeiro and São Paulo stock exchanges—Cr$ million in 1959 (1959–1978)

Year	Value (Cr$)	Year	Value (Cr$)
1959	4.0	1969	65.8
1960	7.7	1970	101.8
1961	6.8	1971	474.2
1962	12.7	1972	285.6
1963	18.2	1973	246.0
1964	12.5	1974	145.4
1965	14.6	1975	223.6
1966	7.8	1976	166.5
1967	10.8	1977	156.7
1968	13.4	1978	158.8

Source Comissão de Valores Mobiliários (CVM); Castro (1979, 50)

financial assets. On the other hand, non-monetary assets increased their participation, rising from 5% in 1961 to 76% in 1981 (Tavares and Carvalheiro 1985, 35)

The stock market thus gained new impetus. Gross issue of shares in relation to GDP rose from 2.6% in 1961 to 4.0% in 1965, 4.4% in 1970, and 6.6% in 1971. Table 2.6 helps to illustrate this growth.

This initial impulse of the stock market had a strong decrease in the year 1971 (as can be seen in the table). Beginning in 1966, issuing of shares multiplied in an unresponsive manner, with no major concern about the operational and allocative efficiency of the process. Both the primary and secondary stock markets saw a significant increase in the volume of transactions, with asset prices also increasing artificially. "Entries without any quality were absorbed voraciously" (Vital 1976, 142), and "the lack of proper regulation favored the disorderly issuing of shares" (Vital 1976, 143). According to Castro (1979, 49), "the accelerated high movement itself represented an example of destabilizing speculation. It was the result of a series of instruments to encourage the transfer of financial resources to the stock exchange concentrated in a short period of time." When the measurement of investors' assessments started to be adapted to the real performance of the shares demand—and consequently of the value of the assets—it was generated the great crisis of 1971. One of the main results of this crisis was the loss of confidence of economic agents in relation to the capital market that had just been established. For Vital (1976, 154), "the rampant speculation that was observed in 1971 and the subsequent disillusionment of investors turned away from the stock market the investors who came to see in the stock market business like game of chance."

According to analysts of the period, this crisis caused "loss of historic opportunity to establish an equity market that is consistent with the required scale and necessities" (Vital 1976, 143). Part of the fortunes that were built in the process of market appreciation was destroyed at the same speed they were built. A large part of the investors turned away from the stock exchange business, due to the trauma caused.

As a result of these processes, companies again only relied on self-financing and loans from public agencies (Vital 1976, 144).

Within this context, in the year 1966 the Brazilian Central Bank authorized the creation of investment banks, which would have the prerogative to give medium- and long-term credits to companies, but also to advise firms to open their capital and place shares on the stock exchange, as well as assisting investors to buy these shares. In addition to these two more general functions, the investment banks would be also responsible:

1. To make investments from two main funding sources: term deposits made by large companies and external on-lending (from international banks and other financial agents);
2. To Underwrite shares (Castro 1979, 47).

Investment banks are considered to be wholesale banks, or "non-monetary" financial institutions, as they may only operate with "third-party resources" (thus with no authorization to receive on-site deposits) (BCB 1966; Teixeira 2000). As Wójcik (2012a, b) shows, they are fundamental instruments for the development of the capital markets, since they are the main agents of the "securitization" of the economies in which they are part, helping companies to open their capital and place shares on stock exchanges, besides creating new financial products for consumption by investors.

The specific regulation on investment banks in Brazil falls under Resolution no. 18 of 1966, which indicates that these banks "are private financial institutions specializing in participation or financing operations, in the medium and long term, to supply fixed and/or working capital, through the application of their own resources and collection, intermediation, and application of third-party resources."

Their operating patterns follow a mix between the North American model of Investment Banks (securities and shares underwriting banks) and the European model of Merchant Banks, or *Banques d'Affaires* (offering long-term financing) (Passos 1973). In short, this new banking provider would have the following functions in the economy:

1. Active operations: loans for financing of fixed and working capital, acquisition of shares, bonds, equities or transferable securities for investment or resale in the capital market; distribution or placing of shares on the market, issues of equities or transferable securities; and the transfer of loans obtained abroad to domestic credit lines, granted by the main private and/or government agencies to provide collateral on loans from within country or from abroad (with consultation of the Central Bank) (BCB 1966; Galvêas 1985).
2. Passive operations: issue, in relation to term deposits, of "certificates of deposit" (CDBs) for the respective depositors; create investment companies or administer mutual investment funds, for the application of capital in a diversified investment portfolio.

In the Brazilian case, two main elements need to be emphasized to understand the role of investment banks in the economy:

1. Even with the efforts made since the Reform of the National Financial System in 1964/65, these banks did not gain significant dimensions (at least compared to their similarity in central world-system countries); commercial banks remain the foundation of the national financial system;
2. After 1988, the major national commercial banks, mainly Bradesco, Itaú, and Banco do Brasil, opened their own "investment banks," which are Banco Bradesco BBI S.A., Banco Itaú BBA S.A., and BB Investimentos S.A. The activities of the large commercial banks with their portfolio of investments hinder the emergence of "pure" investment banks in the national economy.

Thus, some commercial banks in Brazil act as investment banks, although they are not classified as such by the Central Bank (depending on the used criteria). To get an idea of the topology and magnitude of the activities of the investment banks, Table 2.7 shows data on these institutions.

This is the universe of institutions officially recognized as investment banks by the Central Bank of Brazil. Although they are important mainly for stock issuing, and for providing complex financial services (such as advisory services for mergers and acquisitions), their size and weight in the national economy are relatively small, especially if we compare them with their international analogous, or even with commercial/universal banks in Brazil. The total assets of investment banks expressed in the table for 2017 (R$293,916,915,000.00) are almost five times less than the assets of only one commercial bank in the same period (Banco do Brasil, which also had in 2017 R$1,397,970,583,000.00 of assets). The five major commercial banks (Banco do Brasil, Itaú, Caixa Econômica Federal, Bradesco, and Santander) had total assets in the amount of R$5,798,575,247,000.00, that is, 19.7 times more than the assets of the investment banks.

The territorial distribution of investment banks also shows that the Brazilian financial system had up to the 1980s, a more well-regionalized structuration, contrary to its current concentrated and centralized character (Alves 2015). In the first year they were authorized to operate, seven investment banks were created. This number increased to over 41 in 1973, and in 1988, there were 48 institutions, as shown in Table 2.8.

In addition to a significant decrease in the number of investment banks in the country since the 1990s, there has been a growing concentration in the city of São Paulo. These tendencies demonstrate both the consolidation process of investment banks and the primacy enjoyed by the city of São Paulo in the national urban network (Alves 2015).

The capital market again experienced a great boost in the beginning of the 1990s, also due to some new federal policies—directly or indirectly—linked to the sector. Since 1989, with the election of two presidents ideologically tributary to neoliberal ideologies (Fernando Collor de Mello and Fernando Henrique Cardoso), a process of economic liberalization began, based on the flexibilization of regulations of production, commerce, and finance firms (Tavares 1999). São Paulo already had the financial primacy of the national economy, as São Paulo Stock Exchange (Bovespa) started to gather the largest volumes of business, surpassing the Rio de Janeiro Stock

Table 2.7 Brazil—Investment Banks—selected variables (2017)

Financial institution	Property	City	Total assets (R$)	Branches
Banco BTG Pactual S.A.	Private/national	Rio de Janeiro	120,537,328	7
Banco BNP Paribas Brasil S.A.	Foreign	São Paulo	38,940,365	4
Banco de Investimentos Credit Suisse S.A.	Foreign	São Paulo	35,601,335	2
Bank of America Merrill Lynch S.A.	Foreign	São Paulo	20,256,823	1
Banco Crédit Agricole Brasil S.A.	Foreign	São Paulo	18,769,713	1
Goldman Sachs do Brasil S.A.	Foreign	São Paulo	14,506,788	1
Banco Bradesco BBI S.A.	Private/national	São Paulo/Osasco	12,462,775	2
Banco Alfa de Investimento S.A.	Private/national	São Paulo	11,233,741	9
BB—Banco de Investimento S/A	Private/national	Brasilia	7,652,194	1
Haitong Banco de Investimento do Brasil S.A.	Foreign	São Paulo	6,345,911	1
Standard Chartered Bank (Brasil) S.A.	Foreign	São Paulo	2,789,938	2
Banco Itaú BBA S.A.	Private/national	São Paulo	2,538,830	9
HSBC Brasil S.A.—Banco de Investimento	Foreign	São Paulo	1,332,952	1
BR Partners Banco de Investimento S.A.	Private/national	São Paulo	401,645	2
BBVA Brasil Banco de Investimento S.A.	Foreign	São Paulo	136,861	1
Banco Inbursa de Investimentos S.A.	Foreign	São Paulo	115,332	1
Banif Banco de Investimento (Brasil) S.A.	Foreign	São Paulo	88,924	2
Banco Mercantil de Investimentos S.A.	Private/national	Belo horizonte	88,486	1

(continued)

Table 2.7 (continued)

Financial institution	Property	City	Total assets (R$)	Branches
UBS Brasil Banco de Investimento S.A.	Foreign	São Paulo	60,501	2
Banco Porto Real de Investimentos S.A	Private/national	Porto Real	30,403	1
Banco Induscred de Investimento S.A.	Private/national	São Paulo	26,070	1
Totals			293,916,915	52

Source Banco Central do Brasil (2018)

Exchange. This primacy of São Paulo expanded during the period of mass privatization of state-owned companies (between 1991 and 1997), since most privatization deals were carried out on the Bovespa.

This market dynamism was slowed by the successive global financial crises that occurred at the end of the decade (the Asian crisis of 1997, the Russian crisis of 1998, and the Turkish and Argentine crises in 2001). The businesses derived from the privatization process had also depleted its potential for injecting resources into the economy. In the year 2000, the Rio de Janeiro Stock Exchange was almost completely incorporated by Bovespa, and only a small secondary market of public bonds was listed in it. The remaining eight stock exchanges in the country had their operations fully incorporated by the Bovespa. It is also important to mention that in 2008 Bovespa absorbed the BM&F Stock Exchange (the main Brazilian Commodities and Future Exchange, located in the city of São Paulo as well), creating the largest unified securities, commodities, and future business stock market, then "becoming the only stock market for variable income trading in Brazil, accessed by brokerage firms throughout the country" (BM&FBovespa 2017, 31).[6]

As Reinaldo Gonçalves (1999) shows, this liberalization/economic opening had a first moment between 1991 and 1993, which facilitated the exit of capital from foreign firms (transfer of profits and payments for royalties of technological packages), and a second moment from 1995 that "came to eliminate sectoral restrictions on the entry of foreign capital into the Brazilian economy" (Gonçalves 1999, 104). As mentioned,

[6]As shown by the website of B3—the contemporary name of BM&FBovespa—"The integration of BM&F with Bovespa Holdings, approved on May 8, 2008, gave rise to BM&FBOVESPA S.A. Securities, Commodities and Futures Exchange, then already one of the world's largest exchanges in terms of market value. [...] With a totally verticalized and integrated business model, BM&FBOVESPA began operating in the trading and post-trading of equities, bonds and derivatives, publication of quotations, production of market indices, development of systems and software, issuer listing and securities lending, as well as acting as a central securities depository (CSD). Its systems made possible the buying and selling of securities, especially equities, the transfer of market risks via hedging, price arbitrage between markets and/or assets, investment diversification and allocation, and position leveraging, all of which contributed to Brazil's economic growth." (Available at: http://ir.bmfbovespa.com.br/static/enu/perfil-historico.asp?idioma=enu).

Table 2.8 Brazil—Distribution of investment bank headquarters by city (1966–2017)

City	1966		1973		1988		1998		2017	
	No.	%	No.	%	No.	%	No.	%	No.	%
São Paulo (+Osasco)	4	58	13	32	26	56	14	64	17	80
Rio de Janeiro	1	14	14	34	12	25	3	14	1	5
Belo Horizonte	0	0	3	7.5	1	2	2	9	1	5
Porto Alegre	1	14	4	9.6	3	6.5	–	–	–	
Curitiba			2	4.8	1	2	1	4	–	
Salvador	1	14	2	4.8	2	4.5	–	–	–	
Recife	–	–	2	4.8	1	2	–	–	–	
Brasília	–	–	1	2.5	1	2	2	9	1	5
Porto Real	–	–	–	–	–	–	–	–	1	5
Total	7	100	41	100	48	100	22	100	21	100

Source Alves (2015). Banco Central do Brasil (2018)

privatizations meant new "business opportunities" by selling a large group of state-owned enterprises. This wave of privatizations originated at the beginning of the decade (mainly from the edition of the National Privatization Program, Law no. 8031 of 1990) and gained more aggressive contours with the publication of Law no. 9491, of 1997, which also authorized the purchase of entire companies/assets by non-residents in practically all sectors of the economy.

With the Law no. 9491, privatizations were made viable in sectors previously considered strategic for the country, such as the generation and transmission of electricity (and gas), petrochemical sector, transportation (ports, airports, and roads), and especially in the telecommunications sector. The telecommunications sector brought the most results from a financial point of view, mainly due to the sale of state holding company Telebras, which was sold for about R$22 billion, becoming the second largest privatization of the sector in history (exceeding Deutsche Telecom and just behind the Japanese NTT). It is important to emphasize this process, because it was also a powerful instrument of financialization of the national economy, as Rodrigues and Jurgenfeld (2017, 4) note:

> Privatizations boosted the denationalization of the economy, with part of the shares of state-owned enterprises being bought by foreigners, and led to exacerbated financialization, since investment banks and other financial institutions, such as pension funds, were among the buyers in the auctions. In several cases, a few years after buying shares of these companies, financial institutions resold them, confirming their purely speculative objectives with the rapid exit of the business, after measuring significant returns.

Therefore, the 1990s were a period of securitization and internationalization of the economy, as we mentioned in Table 2.1, which depicts the main periods of Brazilian financial and banking development. Since this period, the Brazilian companies themselves beginning to put shares in the international stock market, mainly through the American Depositary Receipts (ADRs) on the New York Stock Exchange (NYSE), "with the objective of capitalizing through the launch of securities abroad" (CVM 2019). A process of rationalization of the Brazilian securities market begun, as companies that were internationalized had to follow a more rigid pattern of organization to adhere to the requirements of the Securities and Exchange Commission (SEC) of the United States, in terms of accounting, transparency, and disclosure standards, the so-called "corporate governance principles."

In 2000, another important event occurred for the national stock market: the promulgation of Central Bank Resolution no. 2,690, which allowed exchange to be organized as joint-stock companies, initiating the process of demutualization of stock exchanges in Brazil (which was already a worldwide process) (Nabarro 2016). The demutualization was completed in 2007, when the Bovespa became a publicly held holding company and made its first public offering of shares.

After the "boom" generated by the privatization negotiations of the 1990s, in 2003 a new market dynamism was started, with many companies being listed on the exchange. As shown by the Securities Comission's own document, during the entire period from 1996 to 2003 only four capital openings occurred, less than one per year; however, between 2003 and 2011, more than 100 new companies were listed on the exchanges (CVM 2019).

The transformation of the BM&FBovespa into a publicly traded company allowed for a greater modernization of its technical systems, in addition to catalyzing its internationalization. BM&FBovespa S.A. also bought the Brazilian Settlement and Custody Company (CBLC—*Câmara Brasileira de Liquidação e Custódia*). As a result of these changes, other financial products were included in its portfolio, such as futures contracts, foreign exchange operations, public and private fixed income securities, auctions, ETFs (exchange-traded funds), carbon credits, among others (Nabarro 2016).

In 2017 BM&FBovespa S.A acquired another financial institution, the Securities Central Custody and Financial Liquidation Registry (CETIP—*Central de Custódia e de Liquidação Financeira de Títulos Privados*), which is also the main institution for dealing with complex securities, such as derivatives. With the acquisition of CETIP, the BM&FBovespa S.A changed its social name to B3, a business name that joins the words, *Brasil, Bolsa,* and *Balcão,* which means *Brazil, Exchange*, and *Over-the-Counter Market*. According to the institutional document of the company:

> [its] main objectives are to manage organized equities, securities, and derivative markets, along with providing registration, central depository, compensation and settlement services as central counterparty to guarantee financial settlement of transactions carried out in its trading environments (B3 2017, 34).

An important factor noted by Oliveira (2010, 96/97) is that even with the recent development of the capital market in Brazil, "companies are heavily dependent on self-financing to finance their growth." In relation to the national average, about 55% of the investments by companies in the country are derived from self-financing, 25% are carried out with the issuing of debts, and only 20% with the stock market. Even large firms—which are assumed to be more modern from an accounting and operational point of view—depend on 48% self-financing, 29% on debt, and only 24% on investment resources from the sale of shares in the primary market. This leads the author to conclude that "companies in Brazil are financially constrained, given the low level of indebtedness (total debt/net worth) and the modest credit-to-GDP ratio" (Oliveira 2010, 97).

Another point highlighted by Oliveira is that the stock market in Brazil remains "highly selective" and "concentrated." The relatively small number of firms that open their capital and sell shares on the stock market (primary market), as well as the concentration of secondary market trading in a small number of firms, are reflections of this relatively small degree of the dynamism of Brazilian capital market (Oliveira 2010, 122).

2.4 The Banking Topology: From Presential to Informational Channels

One of the main forms to understand the financialization of Brazilian territory in the way that we are proposing in this book is the identification of the diffusion of the socio-technical networks that allows the banking system to spread its rationality broadly through the geographic space. For the study of this network, the geographers Santos and Silveira (2001) suggested the concept of "banking topology," which would be the result of the territorial division of labor that credit institutions create in spaces in which they operate.

To understand the banking topology it is necessary, therefore, to start with its typology, i.e., to identify the main characteristics of each of the "units" that constitutes it. Both the regulation institutions (i.e., monetary authorities) and the operator entities of the national financial system (i.e., the monetary and non-monetary firms) have their own internal organization, which is materialized in the territory through a network of units consisting of headquarters, subsidiaries, branches, and channels of provision of basic services/products. Each unit of these networks, therefore, involves a certain level of technical and functional complexity, and develops certain roles determined by the internal division of labor of these institutions (or firms).

The topology of the monetary authorities of national financial system is at the very core of the Brazilian banking topology. The main element of the monetary authority—the Central Bank of Brazil—has a very representative topology of this logic: the headquarter is located in Brasília, where are also all other main regulation institutions of the financial system, including the National Monetary Council (Conselho Monetário Nacional, CMN). But there are other monetary authority entities distributed through some other cities as well, executing functions correlated to those of Central Bank in Brasília. This is the case of the *Comissão de Valores Mobiliários* (CVM)—which corresponds in Brazil to the US Securities and Exchange Commission (SEC), as mentioned—and has its headquarter located in the city of Rio de Janeiro (where is also their board of directors and correlate departments). Another fundamental public autarchy that constitutes the Brazilian monetary authority is the Superintendence of Private Insurance (Superintendência de Seguros Privados, SUSEP) with headquarter in the city of Rio de Janeiro as well, with subsidiaries in the cities of São Paulo, Brasília and Porto Alegre.

The Central Bank of Brazil has a topology organized likewise in regional sub-headquarters, in the main metropolitan areas of the territory (the regional directorates). They are located in Belem (in the state of Pará), Belo Horizonte (Minas Gerais), Curitiba (Parana), Fortaleza (Ceara), Porto Alegre (Rio Grande do Sul), Rio de Janeiro (Rio de Janeiro state), São Paulo (São Paulo state), and Salvador (Bahia). In the headquarter in Brasília are obviously located the bank presidency and the eight board of directors that effectively commands the territorial normative contents related to financial activities and make the decisions that are executed by the internal body of the institution. In the regional directorates are executed the orders from Brasília, in addition to the production of information about the economic and financial dynamics of the regions under their respective jurisdictions (Map 2.3).

If the monetary authorities are basically located in the cities of Brasília and Rio de Janeiro, what is the localization standard of the main financial operators, i.e., of Brazilian banks? In quantitative—and proportional—terms, it was already mentioned that the Brazilian financial system is expressly dominated by "universal" commercial banks. How did the topology of these agents evolve, as well as the distribution of

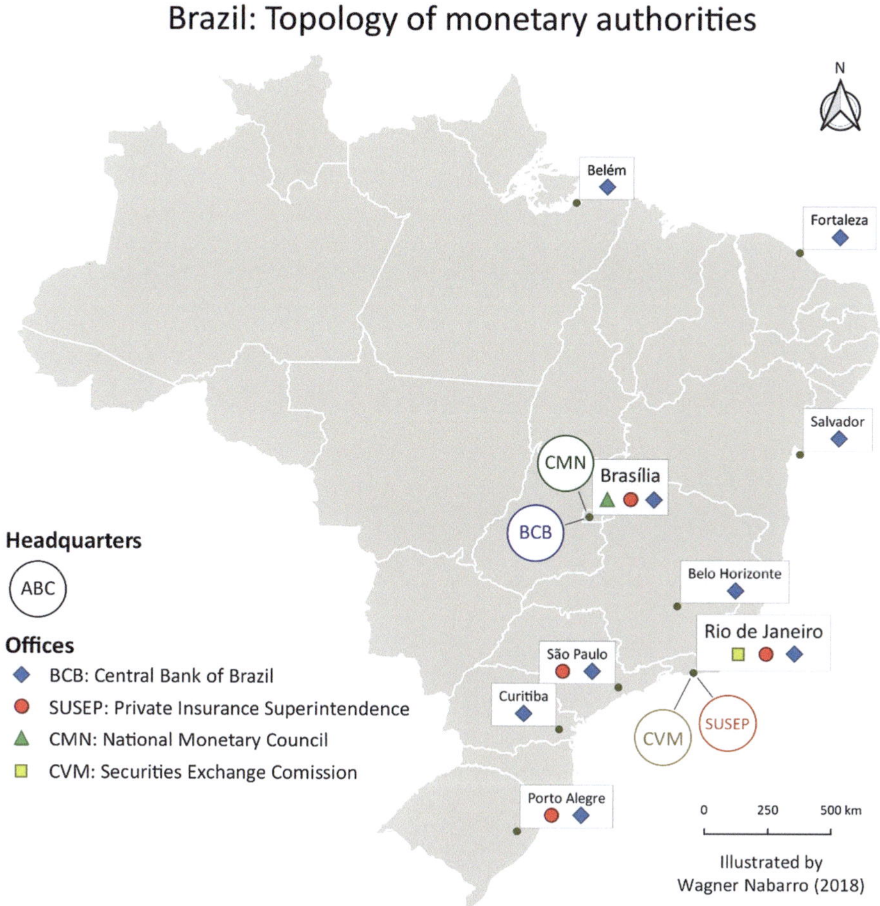

Map 2.3 Brazil—Topology of monetary authorities (2018)

banking wealth in Brazilian territory, after the consecution of Real Plan? Table 2.9 shows this evolution, using data from December 1994 (i.e., the year of the Plan promulgation) up to December 2017, considering three main variables: number of institutions, assets, and number of bank branches in each state.

The table shows the significant reduction of the number of institutions—i.e., commercial banks—in the Brazilian financial system. In the referenced period, the reduction was 117%, number that confirms the expressive process of centralization of capital after the Real Plan. Contrary movement occurs with the two other variables that the table presents: the assets held by the banks grew impressive 1814%, while the number of agencies increased only 20%. São Paulo consolidates itself as the main financial center of the nation in all variables analyzed: if from the point of view of the number of banks the state increased only 1.8% in its importance but in terms of the percentage of assets and banks branches, the state shows all its strength.

Table 2.9 Brazil—Recent evolution of banking assets and topology—in R$ (1994–2017)

	1994		2017		1994		2017		1994		2017	
	No. of banks	%	No. of banks	%	Total assets	%	Total assets	%	No. of branches	%	No. of branches	%
São Paulo	132	50.2	63	52.0	152,204,776	41.7	3,878,562,828	55.4	6,387	35.9	11,429	53.5
Brasília	5	1.9	4	3.3	119,066,749	32.6	2,687,449,612	38.5	5,501	30.8	8,363	39.2
Rio de Janeiro	58	22.0	9	7.5	27,440,320	7.5	170,118,540	2.4	431	2.4	37	0.2
Minas Gerais	13	5.0	6	5.0	20,893,090	5.7	20,521,759	0.3	1,148	6.5	205	0.9
Paraná	5	1.9	2	1.6	15,607,310	4.3	5,525,539	0.1	1,752	9.8	5	0.0
Bahia	6	2.4	2	1.6	10,389,800	2.9	5,159,626	0.1	462	2.5	7	0.0
Rio Grande do Sul	11	4.2	6	5.0	6,732,426	1.9	117,539,265	1.7	585	3.3	554	2.6
Ceará	5	1.9	1	0.9	5,814,079	1.6	54,046,937	0.8	342	1.9	315	1.5
Pernambuco	3	1.2	0	0.0	2,382,238	0.6	0	0.0	159	0.9	0	0.0
Santa Catarina	2	0.9	0	0.0	1,195,587	0.3	0	0.0	256	1.4	0	0.0
Other states	23	8.7	28	23.1	3,326,736	0.9	51,342,531	0.8	779	4.7	430	2.0
Total Brazil	263	100.0	121	100.0	365,053,111	100.0	6,990,266,637	100.0	17,802	100.0	21,345	100.0

Source Central Bank of Brazil (2018). Available at https://www3.bcb.gov.br/ifdata/

The second financial center that grew significantly in the same period was Brasília, federal capital of the country. Except for the variable "number of banks," Brasília appears as the second marketplace of the Brazilian banking system ranking, much ahead of the third main marketplace, Rio de Janeiro. The concentration of the assets in 1994 was 74.3% of totals in São Paulo and Brasília, becoming 93.9% in 2017 (what means an oligopolization of the Brazilian financial space by these two cities).

A last data can be highlighted in the analysis of this commercial bank's topology evolution, some data that are not explicit in the table above (for being aggregated by states). Since 1994, there was a decrease in number of headquarters in more peripheric states of the federation. The data of Central Bank shows that, for the year 1994, practically all capitals of 25 states of the Federation had at least one bank under their command (as already mentioned in item 2.2). In the states with less economic dynamism, these banks were generally a public institution. This was the case in the northern region of the country, where states owned the following public commercial banks: Banco do Estado de Roraima (BANER), Banco do Estado de Rondônia (BERON), Banco do Estado do Pará (BANPARA), Banco do Estado do Amapá (BANAP), Banco do Estado do Amazonas (BEA), and Banco do Estado do Acre (BANACRE). In the northeast region—also known for its lower economic dynamism when compared to the states of the South and Southeast—capitals of smaller states had their own banks: Maceió (capital of State of Alagoas) hosted its state bank (PRODUBAN), São Luis (Maranhão), with Banco do Estado do Maranhão (BEM), João Pessoa (PARAIBAN), and Teresina (Banco do Estado do Piauí, BEP).

More dynamic Northeastern metropolises had also their own private banks, like Salvador (Bahia), Fortaleza (Ceará), and Recife (Pernambuco). Table 2.10 shows all commercial banks—public and private—which had their headquarters in these capitals of the Northeast Region of Brazil, and which were bought and/or extinguished with the advancement of the mentioned liberalization/privatization process that took place in the 1990s.

This more decentralized—and regionalized—structure was a central characteristic of the Brazilian banking topology, not only to the Northeastern states. In 1994, there were headquarters of commercial banks in 25 states of the Federation and in the Federal District (Brasília). With the advancement of the bank consolidation process, in 2017 these headquarters were located only in 10 states (and in Brasilia). The number of banks headquartered outside metropolitan agglomerations (i.e., installed in medium-size cities) also decreased: in 1994 they were 5 (in Uberlandia, Juiz de Fora, Resende, Mossoró, and Novo Hamburgo) and became only 3 in 2017 (installed in Indaiatuba, Ribeirão Preto, and Uberlândia).

Although bank branches have recently begun to lose their primacy as a channel for providing banking services, they have been, since the beginning of the last century, the main element of the banking topology of the Brazilian financial system. Even today, their importance for the banking system is undeniable.

In countries such as Brazil, where a considerable part of the population that hold bank accounts has low income, the bank branches and face-to-face service points continue to be widely used. Between the years 2015 and 2016, "the banking branches had a growth in the number of transactions made, going from 4.4 billion, in

Table 2.10 Brazil—Banks in capitals of Northeastern in the period prior to Real Plan (1994)

	Control	Assets (in 1000 R$)	No. of employees	No. of branches
Recife (State of Pernambuco)				
Banco Banorte	Private	1.615.155	3.581	83
Banco do Estado de Pernambuco	Public	486.041	3.850	56
Banco Mercantil	Private	281.042	795	20
Fortaleza (State of Ceará)				
Banco do Nordeste do Brasil (BNB)	Public	3.811.309	5.465	181
Banco Industrial e Comercial (BIC)	Private	945.316	1.622	38
Banco do Estado do Ceará (BEC)	Public	559.216	2.411	78
Banco Comercial Bancesa	Private	266.138	1.013	29
Banco Fortaleza (BANFORT)	Private	232.100	540	16
Salvador (State of Bahia)				
Banco Econômico	Private	6.610.749	10.688	283
Banco da Bahia	Private	2.242.955	–	–
Banco do Estado da Bahia (BANEB)	Public	1.115.298	5.653	174
Banco de Desenvolvimento da Bahia (DESENBANCO)	Public	339.341	334	1
Banco Bahia	Private	78.641	8	3
BANCIONAL	Private	2.816	17	1

Source Central Bank of Brazil (2018). Available in https://www3.bcb.gov.br/ifdata/legado.html?lang=1

2015, to 5.3 billion in 2016," as shown by on study of the Brazilian Bank Federation (*Federação Brasileira de Bancos*—FEBRABAN 2017). The figures of the last survey of the Brazilian Federation of Banks (FEBRABAN 2018) show that bank branches still account for 42% of financial transactions with money movements, being the largest channel for this type of transaction, followed by points of sales (with 37%) and mobile and Internet banking which together account for 21% of transactions with (digital) money in the Brazilian banking system.

The banking branches continue to be important not only for consumers, but also in respect to the impacts that they generate for the local economy where they are located, once they engender a constant income flow for the entire functionalism employed, which has positive repercussions for local economy. Since its construction—which expenses are around R$ 300,000.00 to R$ 400,000.00, according to data of the Banking Union of Ceará (2009)[7]—up to the execution of expenses with employees and outsourced services, the branches operate as geographic objects that feed the local economy, and the smaller the municipality, the more important is this economic impact.

A second element even more significant that shows the importance in the location of banking branches—mainly in less developed areas—are the external economies that they create and the potential dynamization of local economy that they generate. The external economies are generated mainly for public and private agents that operate locally. The transportation, custody, and use of financial values are services that have a universal demand in any monetized economy. Public agents, such as local governments of municipalities, or subsidiaries from federal and state-level sphere, many times have to bear with the costs—and risks—involved in the availability of resources for the payment of public functionalism (when this is a typical service that could—or should—be provided by banks). Private agents and entrepreneurs, on the other hand, may also have access only to credits for investment—or even for working capital—when using the service provided by managers specialized in credit, which are allocated into bank branches.

In addition of being important in smaller cities, banking branches in larger cities have increasingly become "credit boutiques," specializing only in more complex financial operations. As shows an article from a specialized banking magazine (INFOMONEY 2016), personal contact is necessary, in co-presence, of employees and clients in agencies in the following transactions:

1. Transference of higher values of funds;
2. Loan and financing requirements;
3. Decision-making about investments;
4. Contracting insurance.

In addition to increasing the operating costs of public and private agents, the lack of branches—and especially bank headquarters—in less developed areas also causes severe local credit restrictions and hinders the generation of endogenous regional development processes, as shown by authors such as Porteous (1995, 1999), Alessandrini and Zazzaro (1999), Chick (2006), Amado (2006), among others.

Once these considerations are made about the importance of bank branches in the Brazilian banking topology, it should also be explained the fact that branches are recently being substituted by other service provision channels, much simpler under an operational point of view, that use more electronic currency—or digital money—than bills and/or physical money. A Report at The Economist shows that the main

[7]Available in http://www.bancariosce.org.br/noticias_detalhes.php?cod_secao=1&cod_noticia=7438 (access in Jun/18/2018).

factors that cause the banking branches to be more expensive are the following: 1. Generally, they are installed on "noble" or very busy areas of the cities, making the land or rental costs higher; 2. They require significant expenses with employees and outsourcing companies; 3. The branches are frequently robbed, and therefore have to spend a significant amount with security issues/equipment; 4. "For most of large retail banks, rent payment, equip and contract personnel may easily be responsible for 40–60% of its total operation cost, with computer systems accounting for most of the rest" (The Economist/Carta Capital 2012, 45).

As several authors have highlighted—and also Leyshon and Thrift (1997)—in the behavioral logics of financial agents, it should be also considered the cultural aspects of the agents, as one of the causing factors of the financialization of geographical space. This argument serves to understand the localization of banking branches as well. The selection of the best localization, besides being submitted obviously to objective and strict criteria of geo-marketing (definition of local potential consumer public, their income, number of retired people, public employees at the surroundings, etc.), also considers the "cultural behavior" of clients regarding the consumption of financial services. The article of The Economist/Carta Capital (2012) shows that the localization is also "the first and most important decision when you choose your branch." After choosing the bank branch, although is possible to carry out all other transactions virtually, the clients still feel that their funds remain physically deposited in the original branches in which they have their accounts. Most part of the clients still feel safer with an agency near their residence/work, and part of them require even the presential interaction with bank employees, "to feel that their money is being well taken care of" (The Economist/Carta Capital, *op. cit.*, 46).

The second main type of banking service provision in face-to-face modality in the Brazilian territory, which had a vertiginous growth in the last two decades, was the so-called banking correspondents (*Correspondentes Bancários*). According to the definition of Loureiro, Madeira and Bader (2016, 5), "Banking correspondents are partnerships between a commercial establishment, such as lottery, mail offices or drugstores and a financial institution, in which the commercial establishment, in addition to their main activity, offers services of this institution." It is possible to understand them as "micro-branches" that operate from a computer and an employee contracted by a commercial bank, and that is installed inside a store or commercial/service facility already in operation. In the Brazilian case, the two largest networks of banking correspondents are located inside mail offices network and in the lottery offices network. The first network is controlled by Banco do Brasil and is called *BB Mais*; the second is controlled by Caixa Econômica Federal (which is also the institution liable in Brazil for executing legal betting/gambling), and its name is *Caixa Aqui*.

As shown by Medeiros (2013), the correspondents as legal entities are authorized to operate in the financial system since 1973, when the National Monetary Council publishes the Circular No. 220, allowing commercial banks to contract companies for bill collection and execution of payment orders. But it was only in the 1990 decade that the convergence between the emergence of new information technologies—and a new flexibility in the banking laws—caused a huge growth of the number of bank correspondents in Brazilian territory. Through Resolution No. 2640 from August 1999—after complemented by Resolution 2707 from March 2000—Central Bank of Brazil authorizes unrestricted hiring of banking correspondents by financial operators, to outsource some of their service provision. The range of services that can be done today in bank correspondents is large, and although being relatively simple from an operational point of view, they are extremely important in meeting this kind of demand for financial services. Therefore, the diffusion of correspondents was also a form of simultaneously increasing the capillarity of banking topology and reducing costs for the provision of these services.

The growth of its use, mainly by the commercial banks, is vertiginous. After 1 year of authorization (i.e., in 2000), already existed in Brazilian territory 13,731 correspondents in operation (while the number of agencies in this same year was 16,396). In the year 2004, there were 46,035 correspondents operating, a number almost three times greater than the number of bank branches for the same year (17,260 agencies).

Several are the reasons for this enormous diffusion of correspondents. In first place, the diffusion is due to the significant economy of costs that they provide for the banks, both related with fixed and variable capital used:

1. Once the correspondents are located inside commercial facilities already in operation (mail offices, lottery agencies, drugstores, groceries stores, etc.) and because they operate with relatively simple electronic equipment, they have a very low cost for installation;
2. By employing individuals that are outsourced, and therefore not considered as "bank employees" (i.e., unionized workers and legally within the bank organization), the wages paid are much lower, and the contracts do not guarantee several labor rights achieved by the banking workers along decades of union/political struggle (such as the right to vacations, participation in profits, food and transportation vouchers, etc.).[8]

In this context, national private banks have increasingly incorporated correspondents into their topology for two main reasons: reducing the flow of people within agencies—to diminish costs—and closing branches in less lucrative places (Dias 2017, 389). The banking correspondents easily spread also because the federal government noticed that their usage would be an efficient form for the "universalization of access" to financial services in the Brazilian territory (despite its continental dimensions). In the year 1999, Brazil had 1,679 municipalities without access to banking

[8]In order to justify that they consist of services with a different nature than what occurs in the traditional bank branches, the commercial banks, and their main class association—the Brazilian Bank Federation—FEBRABAN—has been trying to denominate this new form of service provision such as "non-bank correspondents," instead of "bank correspondents" (FEBRABAN 2011). The labor unions, in their turn, show that the correspondents execute operations that are typically banking services, and the way that they are being regulated, generate in banks a "indiscriminate outsourcing in the financial system, including end-activities" (CONTRAF-CUT 2013).

services, and in 2002, all the 5,570 municipalities had at least one bank correspondent (what meant to the federal government this "universalization" of access).[9]

Finally, it should be highlighted that the banking correspondents were also essential for the consecution of one of the most important federal policies for income distribution in national territory, the *Bolsa Família*, as we will see in the next chapter. The correspondents, therefore, were—and are until today—essential for the accomplishment of several policies that affect the lower income population in Brazil, once they are located exactly at the regions and cities in which the lack of economic dynamism does not justify opening bank branches to meet the demands of local population.

Table 2.11 shows the recent evolution of these two central geographic objects of the Brazilian banking topology.

According to the report made by the Brazilian Bank Federation (FEBRABAN 2017, 11), the movement of accounts and services made in banking correspondents, "positively surprised in 2016," reaching at the peak 5.1 billion transactions (when compared to the 3.2 billion registered in 2015, and even greater than the 1.3 billion registered in 2011). According to the document:

> With a differentiated office hour of bank branches and a network with capillarity throughout Brazil, the correspondents allow the consumer that does not use digital resources (Mobile Banking and Internet Banking) to easily pay bills or make small withdrawals (FEBRABAN 2017, 11)

Given this new context, the banking topology being created in Brazil from 1990 decade has two new main features: 1. its broader geographical reach and 2. its growing complexity, mainly because of the new technical systems that became part of its composition. Although bank branches are still being built in Brazil, they definitively are losing their primacy. As shown by several contemporaneous geographers (Santos 1994a, 1996; Martin 1999; Warf 1989, 2017a, b), the information technologies have altered expressively all instances of social life, including the forms of producing and providing banking services. The introduction of information techniques occurred first in the internal processes of banks (back- and front-office activities) to be gradually taken "outside branches," increasing the capacity of the banks to spread out in geographic space (FGV 2010, 226).

It was in 1990 decade that the massive introduction of technical innovations in the Brazilian banking system expanded its capillarity and increased its efficiency. Paradoxically, some of these changes reinforced the importance of face-to-face modality for the provision of services, such as occurred in bank branches and correspondents. Other innovations introduced allowed several services to begin to be delivered remotely and were very economical in respect to the use of workforce by the institution/service provider. Table 2.12 provides an idea of the recent evolution of banking transactions made in Brazil, according to the technical system used.[10]

---

[9]The most recent data released by the Central Bank show that, for 2014, only Pescaria Brava—a small city in the state of Santa Catarina—had no physical/local banking service provider (BANCO CENTRAL DO BRASIL 2015).

[10]It is important to note that these transactions do not involve the movement of money, but rather consultation of balances, request for services, credit cards, etc.

Table 2.11 Brazil—Recent evolution of number of banking branches and correspondents—in 1000 units (2002–2016)

	2002	2005	2010	2011	2012	2013	2014	2015	2016
Bank branches	17	17	19	21	22	23	23	22	22
Correspondents	33	70	165	161	355	375	346	293	277

Sources Central Bank of Brazil and Febraban (several years)

Table 2.12 Brazil—Recent evolution of banking services provision—in billions of transactions (2011–2016)

	2011		2012		2013		2014		2015		2016	
	No.	%	No.	%	No.	%	No.	%	No.	%	No.	%
Mobile banking	0.1	0	0.5	1	1.6	4	4.7	10	11.2	20	21.9	34
Internet banking	12.1	38	13.7	39	16.5	41	18.0	37	17.7	32	14.8	23
ATM	8.3	26	8.8	25	9.2	23	10.2	21	10.0	18	10.0	15
POS	5.1	16	5.7	16	6.4	16	7.2	15	7.8	14	6.6	10
Branches	3.9	12	4.0	11	3.8	10	4.9	10	4.4	8	5.3	8
Banking correspondents	1.3	4	1.4	4	1.3	3	2.3	5	3.2	6	5.1	8
Contact centers	1.4	4	1.5	4	1.5	4	1.5	3	1.4	3	1.3	2
Total	32.2	100	35.6	100	40.3	100	48.8	100	55.7	100	65.0	100

Source Brazilian Bank Federation (FEBRABAN), several years

The mostly automated modalities dismiss the physical presence of the institution on the local of the "purchase" of the service and increase their remote reach. The main types of banking automation that strongly spread in Brazilian territory as from the 1990 decade were as follows:

1. Automatic Teller Machines (ATMs);
2. Contact centers—telephone answering systems;
3. Electronic Transference of Funds (TEFs);
4. Use of magnetic cards in POS terminals (Point of Sale) installed at stores and other commercial facilities (that also make available banking services) (FGV 2010, 227).

The automatic teller machines are probably the most important technology that allowed a greater amplitude of operation for banks in Brazil, as from the decade of 1990. With the installation of these machines that can make automated digital and mechanical operations, the banks enlarged the amplitude and the scope of their operation all over national space, as well as reduced the necessity of hiring workforce (increasing, therefore, internal productivity). It should be emphasized that the ATM also raised the scope of banking services in their time aspect: the clients may have access to several types of services beyond the period in which the banking branches are open daily (in the Brazilian case, between 10 a.m. and 4 p.m.).

Another aspect that can be highlighted regarding the importance of ATMs for the provision of banking services is their relative localization. The ATMs have three types of main localization in Brazil: 1. at the lobby of bank branches, 2. in kiosks installed at public places with major pedestrian flows, and 3. near/inside buildings destined for commerce/services (hypermarkets, supermarkets, stores, shopping centers, etc.). This pattern is also a facilitating factor for banking clients to reach services provided by them (mainly to withdraw money). By the way, the number of transactions that can be performed in ATMs is increasing, and in some of them it is possible to print bank check sheets for immediate use. The great advantage for the banks, which is behind major spreading of ATMs—as well as all other techniques for banking automation—is "self-service": render to the consumers themselves part of the tasks necessary for consumption/execution of service (Jinkings 2002).

A second essential type of technical system that allows remote banking assistance and which use was expressively enlarged by banks in Brazil are the contact centers (previously denominated call centers).[11] The contact centers allow the execution of a series of banking operations, such as consultation of balances, access to banking account information, solicitation of credit cards or duplicate of documents, charges, among others. These centers used software that automatically recognizes the client's

[11]For Almeida (2015, 381), the Contact Centers are a result of the complexification of call centers, and "this new term assigns the places that, in addition to performing traditional services of telemarketing, concentrate also a greater number of information about the consumers (income, periodicity and purchase typology, residence place, age range, education, etc.). The new management methods of information allow manipulation of data banks each time more complex, capable of crossing referencing and defining consumer potential profile."

demand (mainly through the Automated Response Unit—ARU), directing and rationalizing the call/demand. The contact centers are usually located in cities where public authorities make available major tax exemptions, but also where there is large amount of cheaper workforce available (generally in less developed metropolitan regions) (Almeida 2015). It is another important location factor the nonexistence of active unions, which could push up the wages paid. Besides low wages, the work conditions are, usually, extremely consuming and stressful, as shows Jinkings (2002). For the author, the segmented layout of physical environment itself (the disposition in rows of work "cubicles"), the despotic control of the service time (previously settled of 30 s to 2 min for each call), as well as the rigorous controls for task execution, are demonstrations of the increase of instrumental rationalization surrounding the banking work executed by these centers (Jinkings 2002).

In addition to the "internal" characteristics of operation of contact centers, it should be emphasized that due to the convergence that they allow between banking institutions and the support-network of telephony (both fixed and mobile), they made the action of banks in territory practically ubiquitous, at least for these types of services provided by telephone. Currently, according to the National Telecommunications Agency (ANATEL 2018), there are in the country 301.5 million of telephone devices, 43.7 million being fixed, 859.1 thousand prepaid mobile phones, and 257.8 million mobile phones, what makes the contact centers channel available in the entire national territory.

A third important type of channel for banking service provision is the credit and debit card terminals, the POS terminals. These terminals are part of the electronic/digital payment industry having grown considerably in the recent years in Brazil (ABECS 2018). The POS is an automatic payment terminal that reads data from the client's account stored in the magnetic strips or electronic chips of the card and passes on to the financial institutions, which allows to transfer money between accounts of the payer and the establishment that is receiving the payment. The fixed terminal is the most common type of these machines in Brazil, usually located to checkout counters and connected to a fixed telephone line. There is also the portable terminal type called Mobile Point of Sale (MPOS) that operates in a similar form, adding the capacity of the card reader to be taken to the client (which is largely used in bars and restaurants).

According to the analysis made by Fernando Nogueira da Costa et al. (2010), it is possible to identify the operation of these technical objects through the following elements that constitute them: 1. the issuer—or the card administrator company—which is responsible for issuing the credit card, approve transactions, purchase limits, and issue invoices for payment (it is a financial institution represented mainly by banks); 2. the flag or "label" that communicates the transactions between the buyer and the company issuer of the credit card, and that also defines the regulation and policies that the cards under their responsibility shall follow; 3. the processing of transactions by POS terminals is made by the purchasing or accrediting company, i.e., the company that offers the operational base—the sales point terminal—to the accredited; the accrediting company is also responsible for the terminal maintenance and for transmitting data of electronic transactions, depositing afterward the balance in bank account of the accredited; and finally, 4. the consumer, who purchases goods or contracts services upon payment with a credit/debit card, received by the establishment or shopkeeper. As the study also shows, the POS terminals are extremely important once they constitute an efficient electronic payment system, with less costs than the use of bank checks and that promotes customer "loyalty" in their systems:

> They are strategical for the banks […], once the more common is the use of payment cards by their account holders, less will be the withdrawals in cash and greater will be the monetary multiplier. If the commercial chain between the buyers and sellers is constituted between the own bank clients, there will be not leakage of funds from their electronic flow system (Costa et al. 2010, i).

A fourth essential type of communication channel, that constitutes contemporary banking topology as well, but is more recent in the history of the national banking system, is the Internet banking. The first attempt to use Internet as communication channel with the consumers occurred when private Internet providers start to offer access services to network in the 1990 decade. The Internet banking is the direct use of the bank service system as from a personal computer of the client/user. There are several causes that led to a fast and widespread diffusion of Internet banking in the Brazilian banking system; some of them are more general, others linked to the logic of bank operation. In a general point of view, it should be highlighted that the improvement of Internet infrastructures in Brazil was one of these main factors (process initiated in the 1960 decade). Voluptuous investments were recently made by the State for diffusion of a set of backbones and data transmission networks that allow network diffusion (the same thinking serves for the diffusion of POS terminals). The banks also have an unstoppable search for "cost reduction of transaction processing," as shown by Albertin (1999). As time goes by, the "user learning curve of banking services" also become each time more positive, what facilitated the use of digital gadgets by banking customers. Under the costs point of view, it is important to emphasize that with the use of computer networks, the bank "has no expenses with personnel, installation, neither maintenance of terminals. Each user is responsible for his proper microcomputer, through which will access the service" (Lema 1999, 16). With the Internet banking channel, the provision of services may be performed at any place on earth in which a minimally efficient Internet connection is available. This means that the Internet banking use overcomes even the national frontiers, once the service may be executed as from an order provided outside the county of origin of the account holder/user.

The most recent channel for banking service provision is the mobile banking. This new informational channel uses mobile technical devices for its consecution, mainly smartphones (but also tablets). It may be considered as a kind of Internet banking "deepening," but is also more spread, once depends on the use of devices that are extremely mobile and portable. This channel definitively consolidates what is being denominated "digital consumer" of banking services and products: "Nowadays, more than half of banking transactions in the Country are made digitally, reaching 57% of total," reveals the research of the Brazilian Federation of Banks together with Deloitte consultancy (FEBRABAN 2017). The prior technical innovations (mainly ATM and contact centers) did not reduce as significantly the visits to the bank branches. But these portable technical objects (smartphones and tablets), in addition to reducing the number of visits to bank branches, significantly increased the number of transactions made by the client (FEBRABAN 2017). The enormous convenience by allowing to perform operations from anywhere, at any time, is one of the main explanations to understand this efficiency of information technology in the development of banking businesses (and also in the financialization of the geographical space).

Although all different types of channels grow significantly in the Brazilian banking topology—as shown in Table 2.12—the fully digital systems (Internet and mobile banking) grow at much faster rates than the "physical" or presential channels (branches, banking correspondents, and ATMs). As a global trend in bank operation, there is an increase in the use of informational remote channels, which do not involve the direct action of bank employees in contact with costumers. These digital technical systems give enormous capillarity to the action of all types of financial operators and thus make their action virtually ubiquitous.

References

ABECS (2018) Balanço do Setor. 1° trimestre. 23 p. https://www.abecs.org.br/app/webroot/files/media/6/8/1/bfe25fae7a5f502e3a7433c14e8c4.pdf. Accessed on 12 Sept 2018

Albertin AL (1999) Comércio eletrônico: um estudo do setor bancário. Revista de Administração Contemporânea 3(1):47–70

Alessandrini P, Fratianni N, Zazzaro A (2009) The changing geography of banking and finance: the main issues. In: Alessandrini P, Fratianni N, Zazzaro A (eds) The changing geography of banking and finance. Springer, Heidelberg/New York, pp 1–11

Alessandrini P, Zazzaro A (1999) A 'Possibilist' approach to local financial systems and regional development: the Italian experience. In: Martin, R. (ed) Money and the space economy. Wiley, New York, pp 71–92

Alexandre M, Lima GT, Canuto O (2006) Determinantes das decisões locacionais da atividade financeira. Nova Economia 16(2):243–263

Almeida MC (2015) Em outro ponto da rede: desenvolvimento geográfico desigual e o vaivém do capital nas operações de Contact Center. Department of Geography/University of São Paulo (Unpublished PhD thesis)

Alves CJZ (2015) A topologia dos bancos de investimento no Brasil: primazia urbana e formação do complexo corporativo metropolitano de São Paulo. Department of Geography/University of São Paulo, São Paulo (Unpublished Master's dissertation)

Amado AM (2006) Impactos regionais do processo de reestruturação bancária no início dos anos 1990. In: Crocco M, Jaime Jr FG (eds) Moeda e território: uma interpretação da dinâmica regional brasileira. Autêntica, Belo Horizonte, pp. 147–168

ANATEL (2018) Relatórios Consolidados - Indicadores de 2012 a 2015. http://www.anatel.gov.br/dados/2015-02-04-18-43-59. Accessed on 11 April 2018

B3 (2017) Who we are? http://www.b3.com.br/en_us/b3/about/who-we-are/. Accessed on 21 April 2018

Barat J (1978) A Evolução dos Transportes no Brasil. IPEA/IBGE, Rio de Janeiro

Barbosa, F. de H. (2007). Banco Nacional: Jogo de Ponzi, PROER e FCVS. Revista de Economia Política 28(1):97–115

Barros de Castro A (1972) Sete Ensaios sobre a Economia Brasileira (2 vols). Forense Universitária, Rio de Janeiro

BCB. Banco Central do Brasil (1966) Resolução nº 18, de 18 de fevereiro de 1966. https://www.bcb.gov.br/pre/normativos/busca/downloadNormativo.asp?arquivo=/Lists/Normativos/Attachments/40113/Res_0018_v1_O.pdf. Accessed on 10 March 2018

BCB. Banco Central do Brasil (2015) Relatório de Inclusão Financeira No. 3. 167 p. https://www.bcb.gov.br/Nor/relincfin/RIF2015.pdf. Accessed on 14 May 2019

BCB. Banco Central do Brasil (s/d). Bancos Brasileiros após a Resolução 1524, de 21.09.88. http://www.bcb.gov.br/htms/deorf/e88-2000/texto.asp?idpai=RELSFN19882000. Accessed 09 Mar 2018

BCB. Banco Central do Brasil. Exposição de Motivos no. 311 de 23 de agosto de 1995. Disponível em https://www3.bcb.gov.br/sisorf_externo/manual/11-09-020-001.htm. acesso em 11 Apr 2018

BCB. Banco Central do Brasil. Resolução nº 2640, de 25 de agosto de 1999. http://www.bcb.gov.br/pre/normativos/res/1999/pdf/res_2640_v1_O.pdf. Accessed 23 Apr 2018

Beaujeu-Garnier J (1980) Géographie urbaine. Armand Colin, Paris

Beiguelman P (1977) A Formação do Povo no Complexo Cafeeiro. Aspectos Políticos. Editora Pioneira, São Paulo

Carvalho FCC, De Paula LF, Williams J (2015) Banking in Latin America. In: Berger A, Molyneux P, Wilson J (eds) The Oxford handbook of banking, 2nd edn. Oxford University Press, New York, pp 984–1016

Cassiolato JE (1992) A conexão entre usuários e produtores de alta tecnologia: um estudo de caso da automação bancária no brasil. Ensaios FEE 13(1):166–203

de Castro AB, Souza FP (1985) A Economia Brasileira em Marcha Forçada. Paz e Terra, Rio de Janeiro

Castro HOP (1979) Mercado de capitais no Brasil. A evolução recente. In: Castro HOP (ed) Introdução ao mercado de capitais. IBMEC, Rio de Janeiro, pp 37–56

Chavantes AP (2007) Contribuições da reforma bancária de 1921 para estabilidade do mercado bancário brasileiro frente à Crise de 1929. Leituras de Economia Política 12:25–54

Chick V (2006) Bancos e regiões: algumas lições da experiência europeia. In: Crocco M, Jaime Jr FG (eds) Moeda e território: uma interpretação da dinâmica regional brasileira. Autêntica, Belo Horizonte, pp 79–92

Contel FB (2001) Os sistemas de movimento do território brasileiro. In: Santos M, Silveira ML. Brasil. Sociedade e território no início do século XXI. Record, Rio de Janeiro, pp 357–372

Contel FB (2011) Território e finanças. Técnicas, Normas e Topologias Bancárias no Brasil. Annablume, São Paulo

CONTRAF-CUT. Confederação Nacional dos Trabalhadores do Ramo Financeiro. Contraf-CUT e BC discutem PL 4330, MP 615 e correspondentes bancários. 25 Sept 2013. Disponível em http://www.contrafcut.org.br/noticias/contraf-cut-e-bc-discutem-pl-4330-mp-615-e-correspondentes-bancarios-c0df. Acesso em 24 Jan 2017

Correa RL (1989) Concentração bancária e os centros de gestão do território. Revista Brasileira de Geografia 51(2):17–32

Costa Neto Y (2004) Bancos Oficiais no Brasil: Origem e Aspectos de seu Desenvolvimento. Banco Central do Brasil, Brasília, 156 p

Costa FN (2012) Brasil dos bancos. Edusp, São Paulo

Costa FN, Costa CAN, Oliveira GC (2010) Mercado de cartões de pagamentos no brasil. ABECS, São Paulo

CVM (2019) Securities and exchange commission of Brazil. Activities. http://www.cvm.gov.br/subportal_ingles/menu/international/activities.html. Accessed 13 Feb 2019

Dias LC (2017) O correspondente bancário como estratégia de reorganização de redes bancárias e financeiras no Brasil. GEOUSP (Online) 21(2):384–396

Dow S (1999) The stages of banking development and the spatial evolution of financial systems. In: Martin R (ed) Money and the space economy. Wiley, New York, pp 31–48

Evans P (1979) Dependent development: the alliance of multinational, state, and local capital in Brazil. Princeton University Press, Princeton

FEBRABAN. Federação Brasileira de Bancos (2011) Bancarização e Inclusão Financeira no Brasil. FEBRABAN/FELABAN, São Paulo, 50 p

FEBRABAN. Federação Brasileira de Bancos (2017) Pesquisa FEBRABAN de Tecnologia Bancária 2017. FEBRABAN/DELOITTE, São Paulo, 48 p

FEBRABAN. Federação Brasileira de Bancos (2018) Pesquisa FEBRABAN de Tecnologia Bancária 2017. FEBRABAN/DELOITTE, São Paulo, 48 p

FGV (2010) Tecnologia bancária no Brasil. Uma história de conquistas, uma visão de futuro. Fundação Getúlio Vargas, São Paulo

Furtado C ([1967] 1987) Teoria e política do desenvolvimento econômico. Cia. Editora Nacional, São Paulo

Galveas E (1985) Sistema Financeiro e Mercado de Capitais. IBMEC, Rio de Janeiro

Giddens A (1984) The constitution of society. Outline of the theory of structuration. Polity Press, Cambrige

Gonçalves R (1999) Globalização e desnacionalização. Paz e Terra, Rio de Janeiro

Guerim TM (2017) Território e regionalismo bancário: topologia e estratégia geoeconômica do Banestes/ES. Department of Geography/University of São Paulo (Unpublished Master's dissertation)

Hughes, T. P. (1987). The evolution of large technological systems. In: Bijker WE, Hughes TP, Pinch T (eds) The social construction of technological systems. New directions in the sociology and history of technology. MIT Press, Cambridge, pp 51–82

IBGE (2018) Brazilian Institute of Geography and Statistics. Estatísticas da População. https://www.ibge.gov.br/estatisticas-novoportal/sociais/populacao.html. Accessed 10 Jan 2019

INFOMONEY. Existe futuro para as agências bancárias? 23.08.2016. Disponível em http://www.infomoney.com.br/blogs/imoveis/blog-dos-fundos-imobiliarios/post/5488556/existe-futuro-para-agencias-bancarias. acesso em 14 Apr 2018

Jinkings N (2002) Trabalho e resistência na 'fonte misteriosa'. Os bancários no mundo da eletrônica e do dinheiro. Editora Unicamp, Campinas, p 2002

Kayser B (1966) Les divisions de l'espace géographique dans les pays sous-développés. Annales de Géographie 75(412):686–697

Labasse J (1955) Les Capitaux et la Région. Étude Géographique. Librairie Armand Colin/Cahiers de la FNSP, Paris

Labasse J (1974) L'Espace Financier. Analyse Géographique. Armand Colin, Paris

Lagemann E (1985) O Banco pelotense & o sistema financeiro regional. Mercado Aberto, Porto Alegre

Lema MC (1999) Internet Banking no Brasil: Levantamento do Cenário e Análise das Estratégias adotadas pelos bancos de varejo (Unpublished Master's dissertation). Federal University of Rio de Janeiro/UFRJ/COPEEAD

Leyshon A, Thrift N (1997) Money/Space. Geographies of monetary transformation. Routledge, London/New York

Loureiro ER, Madeira GA, Bader FL (2016) Expansão dos Correspondentes Bancários no Brasil: uma análise empírica. Banco Central do Brasil. Trabalhos para Discussão no. 433. 42 p

Martin R (1999) The new economic geography of money. In: Martin R (ed) Money and the space economy. Wiley, New York, pp 1–27

Matias AB (1999) Insucesso de Grandes Bancos Privados Brasileiros de Varejo. School of Economics, Business and Accounting/University of São Paulo, Ribeirão Preto (Unpublished Habilitation thesis)

Medeiros DA (2013) Financeirização do território e circuitos da economia urbana: agentes de crédito, técnicas e normas bancárias. Um exemplo em Alagoas. Departamento de Geografia/Universidade de São Paulo (Unpublished Master's dissertation)

Mendes R (2005) Território e regulação. Espaço geográfico, fonte material e não-formal do direito. Editora Humanitas, São Paulo

Nabarro W (2016) O mercado de capitais no território brasileiro: ascensão da BM&FBovespa e centralidade financeira de São Paulo (SP). Departamento de Geografia/Universidade de São Paulo, São Paulo (Unpublished Master Degree dissertation)
O'Brien R (1992) Global financial integration. The end of geography. The Royal Institute for International Affairs, New York
Oliveira GC (2010) O mercado de capitais brasileiro no período recente: evolução e singularidades. In: Marcolino LC, Carneiro R (eds) Sistema financeiro e desenvolvimento no Brasil: do Plano Real à Crise Financeira. Publisher Brasil e Editora Gráfica Atitude, São Paulo, pp 89–128
Oliveira GPL (1979) Estrutura do sistema financeiro nacional. In: Castro HOP (ed) Introdução ao mercado de capitais. IBMEC, Rio de Janeiro, pp 57–78
Paiva M (2012) BNDES: um banco de história e do futuro. Museu da Pessoa, São Paulo
Passos CF (1973) Estrutura financeira e Desenvolvimento. O caso do Brasil. Atlas, São Paulo
Pires H (1997) Reestruturação Inovativa e Reorganização das Instituições Financeiras do Setor Privado no Brasil. Geo UERJ 2:65–79
Porteous D (1995) The geography of finance. Spatial dimensions of intermediary behaviour. Ashgate Publishing Limited, Aldershot
Porteous D (1999) The development of financial centres: location, information externalities and path dependence. In: Martin R (ed) Money and the space economy. Wiley, New York, pp 95–114
Rangel I ([1957] 1990) Introdução ao desenvolvimento econômico brasileiro, 2nd edn. Bienal, São Paulo
Rangel I (1980) Recursos ociosos e política econômica. Hucitec, São Paulo
Rodrigues CHL, Jurgenfeld VF (2017) Privatizações no Brasil: a desnacionalização e a financeirização (de Collor ao primeiro governo FHC). In: XII Congresso Brasileiro de História Econômica. Niterói, pp 1–33
Saes FAM (1986) Crédito e bancos no desenvolvimento da economia paulista (1850–1930). Instituto de Pesquisa Econômicas/USP, São Paulo
Salviano C Jr (2004) Bancos Estaduais: dos Problemas Crônicos ao Proes. Banco Central do Brasil, Brasília, p 152p
Santos M, Silveira ML (2001) Brasil. Sociedade e território no início do século XXI. Record, Rio de Janeiro
Santos M (1994a) Técnica, Espaço, Tempo. Globalização e meio técnico-científico informacional. Hucitec, São Paulo
Santos M (1994b) Por uma Economia Política da Cidade. O caso de São Paulo. Hucitec, São Paulo
Santos M (1996) A Natureza do Espaço. Técnica e Tempo, Razão e Emoção. Hucitec, São Paulo (Translated to French in La Nature de L'Espace, L'Harmattan, Paris, 1997)
Tavares M, Carvalheiro N (1985) O setor bancário brasileiro: alguns aspectos do crescimento e da concentração. FIPE/USP, São Paulo
Tavares MC (1964) The growth and decline of import substitution in Brazil. Econ Bull Latin Am (CEPAL) IX(1):1–59
Tavares MC (1978) Da substituição de importações ao capitalismo financeiro. Zahar Editores, Rio de Janeiro
Tavares MC (1999) Destruição não criadora. Record, Rio de Janeiro
Teixeira NG (2000) Origem do sistema multibancário brasileiro. Universidade Estadual de Campinas (Coleção Teses), Campinas
The Economist/Carta Capital (2012) Crepúsculo dos Deuses. Reportagem especial sobre o setor bancário internacional. Editora Confiança, São Paulo, pp 41–57
Trigueiros FS (1966) Dinheiro no Brasil. Reper Editora, Rio de Janeiro
Triner G (1999) Banks, regions, and nation in Brazil, 1889–1930. Latin Am Perspect 26(1):129–150
Vidotto CA (2005) Reforma dos bancos federais brasileiros: programa, base doutrinária e afinidades teóricas. Econ Soc 24:1–28
Vital SM (1976) Mercado de capitais. In: Bulhões OG, Castro PR, Vital SM, Waddington A (eds) A Evolução do capitalismo no Brasil. Edições Bloch, Rio de Janeiro, pp 103–174

Waddington AO (1976) Mercado de ações. In: Bulhões OG, Castro PR, Vital SM, Waddington A (eds) A Evolução do capitalismo no Brasil. Edições Bloch, Rio de Janeiro, pp 175–218
Warf B (1989) Telecommunications and the globalization of financial services. Prof Geogr 41(3):257–271
Warf B (2017a) Fiber optics: nervous system of the global economy. In: Warf B (ed) Handbook on geographies of technology. Edward Elgar, Cheltenham/Northampton, pp 113–125
Warf B (2017b) Digitalização, globalização e capital financeiro hipermóvel. Geousp—Espaço e Tempo (Online) 21(2):397–406
Wójcik D (2012a) Where governance fails: advanced business services and the offshore world. Prog Hum Geogr 37(3):330–347
Wójcik D (2012b) The end of investment bank capitalism? An economic geography of financial jobs and power. Econ Geogr 88(4):345–368

Chapter 3
Financialization and Local Scale Dynamisms

Abstract This chapter analyzes some of the specificities of the financialization of Brazilian territory in the local scale. First, the process of credit diffusion to low-income populations is highlighted, showing also the importance of public income distribution policies carried out between 2003 and 2016. The recent development of three main types of "local" financial agents is also explained: credit cooperatives, community banks, and fintechs. It should be emphasized that each of these, in different ways, operates from decentralized—or non-monopolistic—logics. Two of them—the cooperatives and community banks—are not strictly financial, and encourage the engagement of populations in different aspects of their local everyday life, constituting in concrete forms of possible structural changes in financialization as we know it nowadays.

Keywords Place · Bancarization · Indebtedness · Credit cooperatives · Community banks · Fintechs

3.1 Place and Bancarization of Lower Income Population

This new material basis of Brazilian territory, as well as the new banking organization that we described in Chap. 2 has had important consequences related to the increase in the consumption of financial products and services all over the country. As María Laura Silveira (2017, 373) states, "It is there, in places, where we see the increase in the density of financial nexus, the possibilities and the limits offered by credit and the respective increase in consumption." Even the share of the population with lower income started to get in these new financial networks. Instrumental rationality and the temporality of finances begin to be part of the daily lives of populations also in peripheral countries, even in some places further away from the more dynamic areas of these nations. This "financialization of everyday life" in poor countries, as Rankin points out (2013, 549), is mainly linked to the growing role of debt in financing households, but also by the use of financial instruments designed for the most "simple" needs of this population. How can we explain this phenomenon for the Brazilian case, in the context of the financialization on the local scales and the "bancarization" (Crocco et al. 2013; Brown et al. 2013) of lower income individuals?

F. B. Contel, *The Financialization of the Brazilian Territory*, Economic Geography,
https://doi.org/10.1007/978-3-030-40293-8_3

One of the first geographers that studied these themes in Brazil was Milton Santos in his *The Shared Space* ([1975] 1979), book in which the author proposes his "theory of the two circuits of urban economy." For Santos, banks and financial firms, in general terms, can be understood as agents of the "upper circuit of the urban economy," and their action constitutes a "mechanism for drainage out popular savings" (Santos [1975] 1979, 191). He states also that

> The poor […] are those who do not have access, on a regular basis, to goods of ordinary consumption considered as the minimum indispensable in a certain society, and very seldom have access to institutional credit and represent the essential of the clientele of small commercial or craft establishments that offer personal credit at usurious rates. These are, in general, the non-employees and the underemployed, but also of employees who earn very little (Santos [1975] 1979, 38)

More recently, Silveira (2007, 2016, 2017), Montenegro (2010, 2014) and Parserisas (2012) also analyzed the credit phenomenon, based on the theory of the two circuits of the urban economy. For Silveira, one of the main causes of credit diffusion in the poorest groups of the population is related to the global diffusion of neoliberalism (discussed in Chap. 1) and its consequences on state action in peripheral countries. In these countries, the national state no longer provides a range of public services in key social areas such as health, education, and social security. This shrinkage of public services forced the middle and poor classes to consume credit services and products as a way of complementing their lean budget to allow expenditures related to these basic needs that were not necessary previously (Silveira 2007). This gradual decline in social spending, in turn, has allowed a continuous flow of public resources in the form of fiscal allowances, loans, or other types of financing (at low interest rates) to the most modern companies, which make up the "upper circuit of the urban economy." As a logic result, today it is even more difficult to allocate public money to the execution of policies genuinely focused on income distribution and citizenship (Silveira 2017).

For Pike and Polard (2010, 37), "financialization is reconfiguring people's position, financial practices, and articulation within the financial system." In Anglo-Saxon geography, the recent diffusion of financial services and products has been addressed from the perspective of the financial inclusion and exclusion. It was Andrew Leyshon, in a text published on 1995, one of the firsts to systematize this issue in geography. For the author, in fact, the geography relation with "money issues" should debate three main phenomena: (1) the economic geopolitics of finance, (2) the geo-economics of finance, and finally, (3) the study of what the author called the "geography of financial exclusion" (Leyshon 1995a, 534).

In countries such as the United States and Great Britain, the banking restructuring occurred in the 1970 decade ended up significantly modifying the topology of service offers, closing branches and service points exactly in areas of poorer population predominance and/or migrate origin (Afroamerican, Hispanics, Muslims, etc.). For the author, "processes such as these are leading to the deepening of the uneven development in urban areas, and economic and racial dimensions of it are being treated by descriptions such as '*social apartheid*'" (Leyshon, *op. cit.*, 537).

From deepening the financial exclusion, arises at least three processes, observed by Leyshon (1995a) and Thrift and Leyshon (1997):

1. A significant part of the world population is changing their relative position in this new financial economy; from a position structurally explored, this population becomes structurally irrelevant (as also preconizes the thesis of the Spanish sociologist Castells (1996), we would be observing the formation of a "fourth world," even poorer and more excluded than the "third world").
2. New indebtedness spaces appear (i.e., spaces where the credit becomes a central variable for the understanding of local economic and social dynamic).
3. Emergence of alternative forms—or parallel—of organization of activities and financial institutions, such as cooperative and community banks, local exchange systems, alternative currencies, etc.

One of the more consequent analyses in geography about the terms "financial exclusion" and "banking services" was made by Dymski (2007, 267). For him, the financial exclusion "is a central concept in currency and credit geography, which is concentrated on regional localization implications of financial structure for social inequality." The term identifies situations where people have no access to financial products and services (such as bank accounts, credit cards, insurance policies, etc.), and may be also defined as the "condition in which the families or low-income areas are uncapable of obtaining adequate financial services, or can only obtain them at excessively high costs, hindering their capacity to become more prosperous or to avoid trap of debt" (Dymski 2007, 255).

In the origin of studies about financial exclusion are problems associated to racial and ethnic discrimination, and it is important to remember that the exclusion situation may be associated only to different/non-coincident localization of agents and service providers (and not necessarily with any ethnic or economic reason) (Dymski 2005). Obviously, for consisting in a market regulated by eminently capitalist principles, the financial services provision units are distanced from the poorest areas of the territory (whether they are rural or urban). The financial exclusion, however, is directly related to four main factors (isolated, or combined): (1) economic factors (mostly insufficient income of families and/or individuals), (2) ethnical–social factors (race, ethnicity, gender), (3) local factors (absence of banking service providers in communities, cities or entire regions), and (4) regulatory factors (document requirements, minimum age, address proof, etc.) (Dymski 2007, 267 and ss.).

The other important term worked by the author is "bancarization", which we are considering in this book as one of the main bases of the low-intensive financialization. This word gains relevance in economics, geography, and sociology, exactly because of the increase of the banking action scope for populations of poorer countries, once the "low-income financial markets" are being transformed in practically all these countries. According to Dymski (2007, 265), the term "*unbanked* is a new central concept in contemporaneous debates of social sciences about banking consumers. Strictly speaking, a person unbanked is someone that has no bank account in a formal banking institution. Being unbanked is a status."

For Dymski, "bancarization" and "financial inclusion" are not synonymous. If the access and the use of products and services are not suitable for live improvement of families involved, we may not speak properly in "financial inclusion." In this sense, the matter of financial exclusion may be understood also as an exclusion "through the recruitment of low-income families into exploitative credit relationships" (Dymski 2007, 276), i.e., the participation in the financial system, the bancarization, may contribute for the worsening of the economic life of an individual or family. And this exclusion may be perfectly taking place "among those that have banking services and those that don't have," considering that this division "provides little indication about who is financially excluded" (*op. cit.*, 278). A better definition of these phenomena would consider the following classification:

1. Individuals or families *financially included* are those that "have access to a complete set of banking services (…) with tariffs and rates structure that are appropriate considering the transaction costs (for tariffs) and the risks (for loan rates) associated to the supply of these services" (Dymski 2007, 278).
2. Individuals or families *financially explored* are those that "may obtain the banking services only at unfair rates and tariffs (after the proper risk assessment), and that affect the family sustainability" (*op. cit.*, 278/279).
3. Finally, families *financially isolated* (*op. cit.*, 278) are those "located in remote areas, which are excluded from the access to banking services mainly due to their geographic localization" (*op. cit.*, 279).

In the case of peripheral countries of the capitalist system, the rise of the financial inclusion theme has direct relation to experiences associated to microfinances.[1] The pioneer experiences most notable in this field occurred by the action of the Bangladeshi economist Muhammad Yunus (between the middle of 1970 decade and the beginning of 1980). As from the changes introduced in the credit granted by Grameen Bank,[2] multilateral entities—mainly World Bank—sought to promote microfinance diffusion policies in global level, mainly in poor or underdeveloped nations.

For Kraychete (2004), since then, the policies associated to "microfinances" and to "financial inclusion" achieve great relevance, inclusively in Brazil. For the author, the diffusion of microfinances would have the following central concerns (Kraychete 2004, 1/2): (1) election of credit as "one of the central instruments to fight poverty"; (2) concern with "institutionalization" of credit that is practiced by the poorest levels of the population, in order to regulate and make their logic less predatory (substituting therefore the "informal" and more harmful modalities of credit usually practiced);

[1] We will work in this text the definitions proposed by Barone and Sader (2008, p. 1250) for the terms: "microfinances" and "microcredit." The microfinances are the most general phenomenon of diffusion of financial variables for poor population, i.e., "the offer of financial services (productive credit, credit for consumption, savings account, insurance, etc.) for the low-income population that usually has no access to these services by the traditional financial system." The microcredit may be defined as "all financial services for micro-entrepreneurs, excluding the credit for consumption."

[2] Banco Grameen created by him, unlike the "traditional" retail banks, started to use several simple procedures to grant small amount loans to the low-income population of the country, demonstrating to be possible an expressive reduction of costs, a more solidary use of credit and finances, in order to reach "the satisfaction of the necessities of people and assure their well being" (Yunus 2001, p. 174).

and (3) it would initially be liable to the States and NGOs to invest in the constitution of this "formal market" for microfinances.

In the Brazilian case, even if several efforts have been made since the interesting experience of *União Nordestina de Assistência a Pequenas Organizações* (UNO) in 1973 (Feltrin et al. 2009), the microfinances—and the bancarization of population—had a shy economic and geographic expression until the end of the 1990 decade. In the beginning of 2000 decade, almost 30% of the municipalities (1,680 in a total of 5,565) did not have a bank branch, as mentioned. The Central Bank of Brazil estimated that there were around 60 million people with banking accounts in banks in Brazil in 2003 (out of a total population of 182.5 million), while in other countries of Latin America non-financial institutions and credit cooperatives had a greater prominence (Kumar 2004, 12/13). According to Kraychete (2004, 1), the causes of the low rate of bancarization in Brazil in the 1990s—from the point of view of supply of banking products and services—would basically be the following: (1) inadequacy of "regulation pattern" of microfinances, (2) the relative high costs for opening and maintaining the accounts, (3) the low capillarity of the banking system, and (4) the high concentration of banking system (and inhibition of competition in the sector) (Kraychete, *op. cit.*, 2/3).

As the work of Barone and Sader (2008), Feltrin et al. (2009) and Singer (2009) points, this situation starts to change at the end of 1990 decade and gains greater momentum from the two terms of the president of the President Luis Inacio Lula da Silva (2003/2007 and 2007/2011). Since 2003, a set of Plans, Programs, and federal actions induced the Brazilian financial system to perform actions with a clear social-inclusion character. According to Feltrin et al. (2009, p. 22), there was sought a "convergence between microfinances and the traditional financial system, result from the expansion of access to suitable financial services for the diverse population layers efficiently, [and that was] a stage for the achievement of a sustainable and inclusive financial system." These projects would have long-term characteristics, aiming to "rethink the financial inclusion model in Brazil," and "expand the possibility of access to finances" for the entire population (and not for some "groups" or consumer classes) (Feltrin et al., *op. cit.* 22/23).

In this new context, it was performed an expressive combination of alterations in the banking laws and regulation, and several products and services were created and made available by the banks and financial agents, either by virtue of the new federal regulation, or by initiative of the financial institutions themselves. We may aggregate these transformations in four main types: (1) policies directly associated to "microcredit," (2) creation and regulation of the so-called "payroll credit" (*crédito consignado*), (3) institution of new types of simplified and more "popular" bank accounts (simplified savings account and wage/payroll account), and finally, (4) new regulation of the "banking correspondents" (as we saw in Chap. 2). These four sets of changes were responsible for what Levorato (2009, 176) called an "abrupt acceleration of financial inclusion" from 2003. The new products and services available were more "aligned" with the economic conditions of the low-income lenders, and as a rule, do not require formal guarantees that "normal" credits obligated.

It is important to highlight the significance of the creation of payroll loans in this new context of financial inclusion. The three main federal norms that regulate this type of credit in Brazil were as follows:

1. Presidential Decree No. 4.840 dated September 17, 2003, which determined the specific rules for payroll loan for workers registered under the Consolidation of Labor Laws (CLT) (with maximum ceiling of wage commitment of 40%);
2. Decree No. 121 of July 2005, a Normative Instruction of National Institute of Social Security (INSS), which regulated the consigned for retirees and pensioners of the institution (with maximum ceiling of wage commitment of 30%);
3. Presidential Decree No. 6.386 from February 29, 2008, which covers consigned payroll loan for federal public servers (with maximum ceiling of wage commitment of 30–40%) (Levorato 2009, 179).

The two main causes for the enormous growth of this kind of credit (payroll loans)—under the banking sector point of view—were associated to factors that "made the operations feasible" under the commercial point of view: (1) real guarantees for the discharge of the loans (due to the direct discount of the loans in the payroll) and (2) flexible interest rates capable of being charged (except for contracts based on National Institute of Social Security beneficiaries), which, on turn, allowed a reduction in interest rates practiced.[3] According to the vision of bankers, therefore, the payroll loan market was

> very promising and interesting as financial and social inclusion, once promotes not only a productive investment, but other life quality indicators, such as access to durable goods (refrigerators and ovens, for example), better control of finances (with payment of former loans which interest rates were superiors), execution of personal projects, etc. (Levorato, *op. cit.*, 181).

The set of normative changes that allowed the possibility of opening "simplified accounts" (whether "check accounts," "savings accounts," or the so-called "wage accounts") was also very important in this context. Two factors explain this enormous "success" of these new accounts: easy opening (reduced requirement of documents and income prove) and exemption of tariffs (Levorato 2009, 183).[4]

Finally, but not less important, the last element of "financial inclusion" policies from Federal Government that allowed a significant bancarization of the Brazilian population was the possibility of using banking correspondents for provision of simpler services. By their own characteristics—as we saw in Chap. 2—the correspondents have been one of the "most efficient bancarization tools in Brazil" (Levorato

[3]For the author, the interest reduction process with the diffusion of consigned payroll loan occurred, "also due to the guarantees, once the financing debits with salaries, pension or retirement may reduce significantly the default risk, which tend to push down funding rates" (Levorato 2009, 180).

[4]In return for these facilities that were made available by new types of banking accounts, the following conditions were stipulated for their use: the simplified account shall be the only account of the account holder; the maximum limit balance was R$ 1,000.00, at any moment (as well as the limit of deposits); their movement is made only by magnetic/debit card (checks are not offered for its manipulation); finally, the account holder would be entitled to perform only four withdraws, four withdraws of statements, and/or four deposits without charging rates from account holders (Levorato 2009, 184).

2009, 182). The correspondents were important exactly for this better distributed form of expanding access for the population to financial services:

> For a country of continental dimensions, many of its parts of difficult access, with reduced population and/or low income, the correspondents allowed to overcome several obstacles for the expansion of service stations. The main one refers to installation cost at places where the business scale does not pay off such investment (Levorato, *op. cit.*, 182).

The correspondents increased the accessibility of the population to banking services for having been spread in a very diffused topology, as we try to demonstrate in a previous work (Contel 2011). They allowed greater proximity with the consumers (since 2002 all 5,570 municipalities had at least one correspondent to serve their population, as mentioned), but also a new "temporal" condition: with them, the availability of financial services "outside" the regular operation hours of the banking branches became possible (Levorato 2009, 182).

With these Federal public policies of a clearly social and distributive nature, added to the new banking topology of the territory, practically all the economic indicators related to the "real economy"—and not only the financial economy—grew significantly from 2003. Analyzing this evolution in the period from 2006 to 2010, study from FEBRABAN (2011) shows the following changes in the Brazilian economy:

1. A significative growth of Gross Domestic Product (GDP) that went from R$ 2.37 billion to R$ 3.67 billion, a growth of 55.1% in these 5 years analyzed;
2. Increase of income *per capita* from R$ 12,769.00 to R$ 19,016.00 (rise of 48.9%);
3. The unemployment rates that were 8.4% in 2006 become 6.2% in 2010;
4. The Minimum Wage went from US$ 163.55 to US $ 301.78 (rise of 85.5% in dollar).

In addition to these general macroeconomic changes, which increased labor remuneration and consequently the consumption of products and financial services, the strong bancarization would not have been possible if had not been also implemented the policies of social character—especially the *Bolsa Família* Program (PBF)—which allowed the poorer population in the country to have minimum monthly income guaranteed by the Federal Government.

The *Bolsa Família* Program (PBF) was created originally by Provisional Decree No. 132, from October 20, 2003, and definitively institutionalized by Federal Law No. 10.836, from January 9, 2004 (also at the first government of Luis Inacio Lula da Silva). According to Caixa Econômica Federal (CEF)—main public financial institution that manages and distributes the resources of the Program in national territory—the *Bolsa Família* is a "direct income transfer program", focused on families under poverty situation and extremely poor families in the entire Country, in a way that they are capable of overcoming the vulnerability situation and generating conditions for their future emancipation. The *Bolsa Família* allows these families with very low income to receive a benefit in cash monthly, and to receive the resources it is mandatory to the families to keep their school-age children enrolled in the nearest public school of their homes. The management of the Program is decentralized, being

Table 3.1 Brazil—Payment of *Bolsa Família* at the Lottery network of Caixa Econômica Federal (2003–2009)

	Total of payments (in R$)	Payments at the lottery network of CEF	% of payments at Lotteries of CEF
2003	4,986,981	3,764,897	75.49
2004	47,518,163	34,492,687	72.59
2005	80.733,926	59,138,623	73.25
2006	113,899,575	76,162,332	69.50
2007	127,307,177	91,894,431	72.18
2008	126,424,024	89,814,290	71.04
2009	128,202,719	78,285,519	61.06

Source Canton (2010). Lottery Network in Brazil 2010

attributed also to the states and municipalities of the Federation the registration and supervision of beneficiated families. According to the Social Development Ministry (MDS), there were in 2018 around 13.9 million families that benefit from the Program, with the federal expenses around R$ 2.4 billion each month, and the average value of monthly benefit, for each family, of R$ 178.46 (around US$ 47.00).[5]

As also shows the book produced by the Institute for Applied Economic Research (IPEA) and by Caixa Econômica Federal (Canton 2010), a complex system of distribution channels was created by federal authorities in order to deliver the money destinated for social policies like *Bolsa Familia* to people in the proper city where they live. Many Lottery agencies start to open banking correspondents from Caixa Econômica Federal ("*Caixa Aquí*"), and this combination expressively grew the volume of resources of *Bolsa Família* Program that was available by this network of banking correspondents, as shown in Table 3.1.

In addition to the use of banking correspondents from Lotteries, the payment of *Bolsa Família* was also a reason for bancarization of population due to the increase of the banking branches network of Caixa Econômica itself that this policy catalyzed. As shows the numbers of the Central Bank of Brazil, in 2002, Caixa Econômica Federal (CEF) had 1,701 branches in the Brazilian territory; in 2008, this number rises to 2,068; in the year 2012 they are 2,868 and in December 2016 CEF had 3,412 branches in its network (BCB 2017). The growing numbers of banking correspondents of the institution are even more impressive: in 2007, CEF had 13,901 units; in 2012, 36,958 are hired, reaching in 2017 26,807 banking correspondents of CEF in operation scattered throughout the territory (BCB 2017).

[5]The families considered "extremely poor" are those that have monthly income utmost R$ 85.00 per person. The families considered "poor" are those that have a monthly income between R$ 85.01 and R$ 170,00 per person. The poor families can join the program once they have in their composition pregnant woman and child or adolescent from 0 to 17 years old. (Source: http://mds.gov.br/area-de-imprensa/noticias/2018/maio/bolsa-familia-atende-mais-de-13-9-milhoes-de-beneficiarios-em-maio).

This increase in the use of financial instruments did not take place only in Brazil. A research made available by the World Bank entitled Global Financial Inclusion Database (FINDEX) shows in percentile terms interesting data to measure the bancarization of population in Latin America.

It can be emphasized, first, the relative position of Brazil in Latin American context. The country overcomes the other neighbors in all indexes highlighted (bank accounts, credit accounts, and debits cards). Another aspect of the financialization of these territories that the data explain is also the enormous inequality existing between more "bancarized" countries and the less "bancarized" ones, as observed in the study of FELABAN from 2007 as well (Rojas-Suárez 2007). While in Brazil around 68.1% of the people in the selected age have a bank account, in Peru this index is 29% and in Paraguay corresponds to 21.7% (around one-third in a direct comparison with Brazil). A similar thinking serves for the owners of credit cards in these populations as well: Uruguay, with greater use index—39.8%—stands out in respect to 5.7% of Ecuador and the 6.2% of Bolivia. These differences are more acute when we compare all indexes from the countries selected in Latin America with the United States, as the Table 3.2 shows. Finally, we should remind that, parallel with these profound differences, the last report of World Bank over the theme shows that bancarization of populations in Latin America continues to grow (Demirguc-Kunt et al. 2015).

The financialization of Brazilian territory, in respect to the diffusion of the financial variables focused on the lower income population consumption, was also a reality

Table 3.2 Latin America—Evolution of number of bank accounts and credit and debit cards—selected countries (in % of population with more than 15 years) (2011–2014)

	Bank accounts		Credit cards		Debit cards	
	2011	2014	2011	2014	2011	2014
Brazil	55.9	68.1	29.2	32.0	41.2	59.2
Chile	42.2	63.3	22.8	28.1	25.8	54.1
Venezuela	44.1	57.0	10.4	21.5	35.1	49.6
Argentina	33.1	50.2	21.9	26.6	29.8	44.2
Uruguay	23.5	45.6	27.1	39.8	16.4	37.7
Bolivia	28.0	41.8	4.1	6.2	12.8	23.1
Ecuador	36.7	46.2	10.2	5.7	17.1	25.6
Mexico	27.4	39.1	13.0	17.8	22.3	26.8
Colombia	30.4	39.0	10.2	13.7	22.7	30.0
Peru	20.5	29.0	10.0	11.7	14.1	21.4
Paraguay	21.7	21.7	9.0	9.0	11.3	11.3
Latin America	39.3	51.4	18.2	21.6	28.9	40.4
United States	88.0	93.6	61.9	60.1	71.8	76.2
World	50.6	61.5	14.9	17.6	30.5	40.1

Source World Data Bank. Global Findex (Global Financial Inclusion Database) 2016

as from the beginning of the century. Under the point of view of a geographic analysis, it is possible to say that the policies that changed the bank regulation of the territory were a fundamental condition for the financialization of national space, and consequent bancarization of the lower income population. In this process, we highlight also the significance of the federal social public policies to this bancarization, but also the active role of the new banking topology to this process.

The process of diffusion of financial products and services for this population obviously did not only bring improvements to their living conditions. The indebtedness and insolvency of the families that took credit and loans is currently one of the main adverse consequences of bancarization. A recent survey by the National Confederation of Commerce (Confederação Nacional do Comércio), made from a sample research with families living in metropolitan areas, reveals that this popular indebtedness is high, and has recently grown in Brazil.

As can be seen in the Table 3.3, just over 60% of Brazilian households currently have some type of indebtedness, and about 25% have past due accounts. The number that most catches the attention, however, is the third one: more than 10% of families are not able to afford their debts (CNC 2018), which characterizes a full default situation. Two other types of data draw much attention in the survey: in the year 2017, 76.7% of these families had debts on their credit cards, by far the largest percentage cause of indebtedness. The second main form of indebtedness is payment bills—*carnets* (15.7%)—used mainly to purchase consumer goods such as clothes and household appliances; the third is personal credit (obtained in banks or other financial institutions) (10.3%); the fourth form is vehicles financing (10.2%); and fifth leading cause of indebtedness is real estate financing (CNC 2018, 3). The general average of the commitment of these families to the payment of debts in 2017 was 30.1% of their domestic budget (CNC 2018, 4).

Added to these negative effects, it should also be remembered that the positive side of the process (the more social and inclusive nature of federal public policies that were going on) was strongly attacked since the year 2016 when the President Dilma Rousseff suffered a parliamentary coup that resulted in her impeachment (Dos Santos 2017). The timid—but steady—process of income distribution that was underway in the governments of the Workers' Party (Partido dos Trabalhadores) was definitely discontinued. Since then, neoliberal policies, more concerned with financial market issues than with the economic and social development of the nation, began to be

Table 3.3 Brazil—Recent evolution of indebted or defaulted households—in % (2010–2017)

	2010	2011	2012	2013	2014	2015	2016	2017
Indebted families	59.1	62.2	58.3	62.5	61.9	61.1	60.2	60.8
Families with overdue accounts	24.9	22.9	21.4	21.2	19.4	20.9	24.2	25.4
Families unable to pay overdue debts	8.9	8.0	7.1	6.9	6.3	7.7	9.2	10.2

Source Pesquisa CNC (2018). Endividamento e inadimplência do Consumidor

implemented again, increasing the economic and social crisis and moving away the country from the possibility of a more just and egalitarian development.

The next three subsections of this chapter seek to show other forms of low-intensive financialization of the territory, all of them indicating the possibility of less instrumental uses of finance in Brazil. This is the case of the *credit cooperatives* (or even *cooperative banks)* and *fintechs* that points to the possibility of non-monopolistic uses of finance, while the *community banks* seems to be a popular alternative for financialization, with frank empowerment of the communities involved in the creation of these banks. These new forms of financial organization on a more solidarity basis "can therefore be understood as the will of certain economic actors to propose a concrete alternative to 'financialized capitalism'" (Lagoarde-Segot 2015, 56).

3.2 Credit Cooperatives: Finances Between Local Solidarity and Centralization

The study of credit cooperatives deserves to be made by the very nature of this type of financial organization. Born in the beginning of nineteenth century in Europe, they bring since their origin elements of the so-called "utopic socialism" of Saint-Simon, Charles Fourier, and Robert Owen, authors of which "the cooperativism received fundamental inspiration" (Singer 2002, 38). As shows the same author, the cooperatives emerge to help the poorer population, by means of philanthropic actions, what makes their activity extremely important for the social context in which they are installed (Singer 2002, 66). Recently, Santos et al. (2017, p. 33) point also that the cooperatives are interesting as financial agents for presenting three basic characteristics of their operation:

1. The mutual aid between the cooperative members;
2. The self-responsibility in management;
3. Democratic self-determination of its operation.

Since then, several historic experiences were made, successful or unsuccessful, but that seem to indicate two essential characteristics of its operation, which invites us to consider them in this work about the financialization of Brazilian territory: the solidarity principle that is on the basis of its operation and the more local character of its action. Several authors in Brazil denominate these more local forms of financialization of the territory of "proximity finance" (*finanças de proximidade*) (Abramovay 2004; Burigo 2006; Silva 2017).

The eminently local character of credit cooperative is an attribute of their nature, since the creation of the first associations that had this name. Friedrich Wilhelm Raiffeisen (1818–1888) is responsible for the creation, in the year 1872, of *Rheinische Landwirtschaftliche Genossenschaftsbank*, which was the first "regional bank" that served as a kind of "central bank of credit cooperatives in the region," which allowed to maintain the decentralized and solidary character of singular cooperative action in Reno region in Germany (Singer 2002, 65).

Highlighting the local character of cooperatives is important once it catalyzes and guarantees a trust system between the actors involved, due to the fact that all participate in the same community of interests (whether of political, economic, social, and/or cultural interests). As shown by Singer (2002, 68/69), "Once the members of the cooperative know each other, they are neighbors and operate in the same branch, their endorsement is more valuable than the technical opinion based on a standardized set of information."

The document produced by the Central Bank of Brazil (BCB) and the National Cooperative Learning Service (SNAC) (BCB/SNAC 2016) also shows the reasons for which it may consider the cooperatives as a more horizontal model of service provision and "fitted" into their local realities:

1. All services consumed (loans, payment of bills, savings accounts) are a result of the local necessities and are "reinvested" in the same places;
2. These same financial services are made at "differentiated prices," once the cooperate members themselves are the "owners" of the "business," "which gives them back the surplus of operations performed, and—through it—retain financial resources at the places where they live, benefiting local economy";
3. According to the document analysis, the cooperated, therefore, are not simply clients, they are also "owners of a credit cooperative as from the moment in which they purchase a share of the capital" (BCB/SNAC 2016, 18).

This local character that is one of the basic characteristics of the cooperatives, although, is not a "general" rule. The evolution of cooperatives in Brazil and around the world, in many cases, followed a distancing path from cooperatives of their local origin context. The most dynamic credit cooperatives were little by the beginning of their operations, and some of them soon increased in size and sought to gain scale economies when "merged" into credit confederations (then acting almost like a commercial bank). According to the Central Bank of Brazil, in the Brazilian case, a consolidation process started on the 2000 decade, when the cooperatives start to seek vigorously a greater operational efficiency, which in turn led some of them to merge, and this process provided significant scale economies for the cooperatives (BCB 2016, 10). Under a more general point of view, as shows Singer (2002), the confederations of cooperatives may be fundamental for the entire system of cooperatives, both for reasons of already mentioned "efficacy" and/or "scale economies"—which approximate their performance of a more capitalist logic—but also for diminishing risks for the system as a whole.

This is a worldwide phenomenon. In all countries, cooperatives tend to become organized in syndicates—or confederations—that aggregate great amount of units, in order to constitute entities of a greater complexity level, "to execute in common and in greater scale the economic and assistance services of their interest, seeking to act in an integrated, standardized shared form, with less costs, more reliability and efficiency" (BCB 2016, 58). According to the study of the Central Bank of Brazil, what varies, from country to country, is the amount of systemic levels—two or three—and the option for a more or less vertical structure (i.e., with higher or lower hierarchy level, centralization, and standardization between the parties) (BCB 2016, 58).

For the internal operation of cooperatives, therefore, the increase of their size brings several advantages. The confederation of cooperatives may, for example, mobilize funds with other commercial banks at much cheaper prices than those of singular cooperatives. Cooperatives associated in greater scales (interregional or national) may irrigate with resources areas of the territory/region in reference that eventually may have suffered any problem or conjunctural crises (bad crops, natural accidents, problems in infrastructures), which would not have conditions to face these crises if were singular/local cooperatives. In other words, they have better conditions to face eventual unbalance between surplus areas and deficit areas in credit provision (whatever is the reason) (Singer 2002, 69).

This gain in scale, in turn, also shows its negative aspects. Burigo (2006) suggests that most of the cooperatives that adopt a verticalized organization under the administrative point of view end up impressing in their operation an instrumental rationality, aiming to reach more significative clients (under the economic point of view), and distancing from more common preoccupations in smaller cooperatives, mainly those linked to smaller dimension cooperate members (and the low-income population in general) (Burigo 2006, 323).

The evolution of credit cooperativism in Brazil, unlike the Fintechs and Community banks, is a long history. The emergence of this financial activity is associated to the action of the Swiss priest Theodor Amstad, who emigrated to Brazil in the year 1855, and that in 1902 created the *Sociedade Cooperativa Caixa de Economia e Empréstimos de Nova Petrópolis* (Soares and Sobrinho 2008). Within his religious functions (he was responsible for the catechization of the city and the region of Nova Petrópolis, in the Rio Grande do Sul state), the priest was mainly responsible for traveling among German colonies installed there. During this travel, he could identify a set of similar problems that the crops in the countryside of the territory suffered, such as lack of basic infrastructures (transportation, energy), lack of services such as education, health and—also obviously—financial services (BCB/SNAC 2016, 17).

Because of his strong operation in the region with the diffusion of cooperativism, he assisted in the foundation of other 37 cooperatives, but also other forms of associativism and institutions of public services, such as agriculture unions, hospitals, rest homes, schools, magazines, churches, and new agricultural colonies (BCB/SNAC 2016, 17). The main result of its action associated to cooperativism is Sicredi Foundation (current Sicredi Pioneira), which completed 114 years in 2016, and has a history in financing fundamental goods for communities in which the cooperative operated (financing mainly the construction of houses and purchase of land for agricultural production) (BCB/SNAC 2016, 17). Currently, Sicredi Pioneira is the eighth largest financial cooperative of the country, has around R$ 1.3 billion in assets, and counts with 102,364 associate members. According to the document, the cooperative became a type of "brand" in the city of Nova Petrópolis, contributing for financing its growth and its knowledge of the region (BCB/SNAC 2016, 17).

Having its initial development much associated to the countryside and places without adequate conditions for living and producing, the cooperativism in the first decades of the past century in Brazil developed with a strong political and claimant character, such as unions, social movements, and other similar associations. The first initiatives of cooperativism had two major influences in their genesis:

1. The Raiffeisen (from Germany) and Luzzatti (Italy) models, implemented in the respective areas notable for their German and Italian immigration (in the South states of the country), as well as the Desjardins model of "mutual credit," of Canadian origin;
2. An "eclectic set" of cooperatives that multiplied in the period, mainly in the South and Southeast states, for both agriculture credit, but also popular cooperatives of urban credit, cooperatives of employees in companies (public or private), work cooperatives of a specific class/profession, in addition to "school cooperatives" (Pinho 2004, 14/15; Palhares 2004, 45).

In this period, the main laws that governed the entities were the Decree 22.239 from 1932 and afterward the Decree No. 581 from 1938—both created in the government of President Getúlio Vargas—which maintained the Law spirit under Rochdale principles, more concerned with the associativity and democratic character of organizations than with its efficacy and profitability. These territorial normative contents prohibited at that time all possibilities of transforming the cooperatives in "anonymous societies" (Palhares 2004, 52).[6]

In November 1958, by request of the Superintendency of Currency and Credit (SUMOC), the Agriculture Ministry edited Ordinance No. 1.079, that suspended new registration of credit cooperatives, situation that would be ratified in November 1962, upon the edition of Decree No. 1.503, from the then Council of Ministers. For Soares and Sobrinho (2008, 70), "As everybody knows, this was a period of political turbulence, which culminated in the military coup of 1964, when there was fear about every type of community-based organization, even more in the rural environment, main focus of cooperativism." In the beginning of 1960 decade, cooperativism was seen as a type of organization directly associated to leftist and/or libertarian ideologies (or even "communism").

The result was the retraction of the sector that, at the end of the 50s and beginning of 60s, experienced a fall in the number of institutions (Soares and Sobrinho 2008, 71). Therefore, the cooperatives entered into a type of "ostracism" in the 1960 and 70 decades due to the military dictatorship, which restrained any kind of association of more social or community-based character (mainly in rural areas, where the social movements were also powerful). In this context, as from the year 1967, the creation of new cooperatives was prohibited in Brazil (Kumar 2004, 178). The main event,

[6]The cooperative of Rochdale Equitable Pioneers, created in 1844 in the English city of the same name, is considered the "mother of all cooperatives" (Singer 2002, 39). The principles set forth therein ended up becoming "universal principles" of cooperativism: (1) each cooperative member would be entitled to one vote, independent of how much invested in it; (2) the number of cooperative participants would always be open, being possible to accept all those wishing to associate to the cooperative (Singer 2002, 39).

of positive character, which occurred in this period was the creation of Brazilian Cooperative Organization (*Organização das Cooperativas Brasileiras*—OCB), on December 2, 1969, in an attempt to reorganize the cooperatives (OCB, *apud* Burigo 2006), but in a more controlled way by the federal government.

Under the point of view of the regulation of the activities, the cooperatives are not immune to alterations brought by the already mentioned Reform of Financial System of 1964/65. The Law No. 4.595 defined the cooperatives as financial institutions, ending a regulatory conflict existing between the Agriculture Ministry and the Superintendence of Currency and Credit (that disputed the regulation of the activity). The other federal official norms that were central in this period, regarding the credit cooperatives, are as follows:

1. Resolution No. 11 of CMN from 1965, which "defined the activities allowed and the criteria for authorizations, inclusively in respect to the necessity of renovation of already existing cooperatives, upon the corroboration of the previous register in the Ministry of Agriculture" (Soares and Sobrinho 2008, 71);
2. Resolution No. 27, from June 1966, that allowed credit cooperatives to receive non-cash deposits and only from associate members (Soares and Sobrinho 2008, 71);
3. Law No. 5.764 from 1971 (Cooperative Law) that determined the biannual distribution to associate members from eventual existing surplus, Law that "finally came to offer legal guides to the system as a whole" (Soares and Sobrinho, *op. cit*, p. 71). The edition of Law No. 5.764, in 1971, when established the pyramidal structure, made important steps seeking a better organization of the sector (*idem*, 108).

The cooperativism of credit reappears, still shy, in the 1980 decade as from the resumption of agriculture cooperativism (Burigo 2006, 322). Already in this period, as shows Diva Benevides Pinho (2004, 14), the cooperativism grew much more by the inaction of the State and by the selectivity of commercial banks than by the impulse of the proper sector (they acted in "spaces not disputed by the financial and capitalist banking system"). After the Constitution of 1988, the cooperatives began to have two main objectives: (1) transform into financial institutions to meet the growing credit demands by the small- and medium-size rural producers and (2) obtain the permission of monetary authorities to create cooperative banks (controlled by the proper cooperatives), to gain scale and efficacy in their action (Pinho, *op. cit.*, 15).

Currently, according to a recent study of Central Bank of Brazil, the credit cooperatives may be defined as follows:

> Credit cooperatives are financial institutions that primarily destine, by mutuality, the provision of financial services to their associate members, being assured to them the access to instruments from the financial market. To capture resources and to grant credits and guarantees are restricted to associate members, except for the operations made with other financial institutions and the resources obtained from legal entities, eventually, at favorable rates or free of remuneration (BCB 2016, p. 8)

Regarding their contemporary formal organization, the credit cooperatives in Brazil are divided into three main types: (1) the singular cooperatives—that also maintain the eminent local character in their operation—(2) the central cooperatives or "federations" of cooperatives, and (3) the cooperative confederations.

The singular cooperatives are constituted by the minimum number of 20 individuals, being exceptionally allowed the admission of legal entities that have as purpose economic activities equal—or correlated—to individuals, or non-profit activities. Characterized by the direct provision of services to associate members, each associate will be entitled to only one vote, whatever their number of shares, forbidden to distribute any type of benefit to the shares of capital (BCB 2016, 9).

The central cooperatives—or Federations of cooperatives—present a greater complexity level. They are constituted by, at least, three singular cooperatives, entitled, exceptionally, to admit individual associates. The purpose is to organize, in common and greater scale, the economic and assistant services of interest of the branch offices, integrating and instructing their activities, as well as facilitating the reciprocal use of services (BCB 2016, 8).

Finally, according to the current legislation, also operates in Brazil the Confederations of cooperatives. These institutions consist of, at least, three federations or central cooperatives, of the same or different modalities. The confederation of cooperatives has as purpose to instruct and coordinate the activities of branch offices, when the size of the undertakings transcend the scope of capacity (or convenience) of operation of the centrals or federations (BCB 2016, 8). The four confederations of credit cooperatives currently existing in Brazil are as follows:

1. CRESOL—Confederação Nacional das Cooperativas Centrais de Crédito e Economia Familiar e Solidária—Cresol (headquartered at Francisco Beltrão, Parana state);
2. SICOOB—Confederação Nacional das Cooperativas do Sicoob (headquartered in Brasília);
3. SICREDI—Confederação Interestadual das Cooperativas Ligadas ao Sicredi;
4. UNICRED—Confederação Nacional das Cooperativas Centrais.

The central of cooperatives (federations) are 35 while the singular cooperatives—or of "1st degree"—are 967 (BCB 2018, p. 8). Considering the complexity achieved by the sector of credit cooperatives in Brazil, there are already two cooperative banks in activity: Banco Cooperativo do Brasil S/A—Bancoob—and Banco Cooperativo Sicredi S. A.—Bansicredi.

It should be highlighted that the cooperatives are also a way to provide financial services in predominantly agricultural areas of the Brazilian territory, distanced from major urban centers, areas where the commercial banking system has little—or none—insertion. It is in these areas that appear the interesting phenomenon of the *solidarity credit cooperatives*. This type of pioneer cooperatives was created in the state of Santa Catarina, which already had a relatively long history of cooperative traditions on field. Likewise, in the state of Paraná emerged the credit cooperatives of solidary integration that ended up being the basis of the creation of the already mentioned Cresol System, as well as cooperatives linked to Association for Support Cooperatives for Family Economy (Ascoob) founded in the countryside of Bahia state (Burigo 2006, 324).

After these experiences emerged, the first solidarity credit cooperatives were implemented in Brazil. The main institutionalization of this type of cooperatives occurred with the creation of the National Association of Credit Cooperativism of Family and Solidary Economy (Ancosol), on the year 2004, as from the union of the following cooperatives:

1. Cooperativa Central de Crédito e Economia Solidária (Ecosol);
2. Cooperativa Central de Crédito Rural com Interação Solidária (Cresol);
3. Associação das Cooperativas de Apoio à Economia Familiar (Ascoob);
4. Cooperativa de Crédito Rural dos Pequenos Agricultores e da Reforma Agrária (Crehnor);
5. Cooperativa de Crédito Rural de Itapipoca—CE (Cocredi);
6. Cooperativa de Crédito Rural de Desenvolvimento Solidário de Iúna-ES (Credsol);
7. Cooperativa de Crédito Rural da Agricultura Familiar com Interação Solidária de Jataí—GO (Credijat) (Burigo 2006, 324/325).

In addition to this institutionalization of solidarity credit cooperatives, other initiatives are being taken to expand the access of credit to less developed areas in Brazilian territory, (mainly agricultural areas). As shown by Santos et al. (2017) since the 1990 decade has been arising a set of agricultural cooperative initiatives more directly focused on assisting small family producers, which have difficulties to guarantee a production in scale, as well as to regularly produce their crops (what difficult their insertion in circuits for production commercialization). These family producers consist mostly of people with less formal education and "low investment capacity" (Santos et al. 2017, 32)

By promoting access to credit for this type of small farmers—that therefore join the cooperatives—a series of new processes become part of their work routine. Among the main advantages earned by the cooperatives is the economic development arising from the assistance to the insertion of family products in the market, as well as by the proper increase of social relations based on solidarity and in mutual aid (Santos et al. 2017, 32). This insertion, on its turn, contributes to all that become part of the cooperative to increment their productive capacity. As shown by the authors, these cooperatives "have an important role in the improvement of income distribution in rural area, once they may promote the aggregation of value to agriculture products and increase the bargaining power of rural producer in markets relatively imperfect" (Santos et al. 2017, 33). This study, done in the state of Goiás, was based on interviews and questionnaires in five agriculture cooperatives situated there, and all producers were specialized in production/commercialization of milk and soy, based on family farming. Among the main benefits that the cooperatives provide to their associates would be as follows:

1. Feasible commercial alternative, once they promote and facilitate markets access for producers. In addition to the better prices paid by the cooperative, other practical advantages earned by the cooperative members are the greater sales scale and the "elimination of the commercial intermediaries" (Santos et al. 2017, 38).

2. "They provide for regular purchase of farmers' production and guarantee the payment of products traded, fulfilling the fundamental role of creating competition in local markets and avoiding the fall of prices paid to producers" (Santos et al. 2017, 38/39).
3. The cooperatives internalize the profits earned, opposed to the use of intermediaries. The surpluses achieved with the commercialization has two main destines:

   a. Payment of better prices to cooperate members;
   b. Improvement of proper cooperative.

4. The fact that the directors of cooperatives themselves are also producers/cooperative members, and live together with the same types of problems, increases the "social involvement" of the group (Santos et al. 2017, 39).

To get a more systematic understanding of the relative importance of credit cooperatives in the context of the Brazilian financial system, the Table 3.4 shows data on their assets, credit portfolio, and number of branches and service points.

The Table 3.4 shows, once again, the disproportionate weight of commercial banks in Brazil, which hold almost 84% of total assets, almost 85% of loans granted and about 96% of branch network in the territory. In terms of the quantitative significance of cooperatives—both singulars and confederations—their numbers are relevant, but in relative terms the financial values have a low participation in the financial system: credit cooperatives had only 3.25% of the assets of financial system and 3.14% of the credit operations under their control.

The main highlight, however, concerns the physical presence of the points of service of the singular credit cooperatives in the Brazilian territory. In 2018, this type of cooperative had almost half of the points of service in this topology, a figure

Table 3.4 Brazil—Credit cooperatives and other financial institutions—selected indicators (2018)

Institution type	Total Assets (R$)	%	Credit operations (R$)	%	Nbr. Branches	%	Nbr. service points	%
Commercial banks	7,556,21,821	83.55	2,988,813,340	84.50	21,372	96.03	5,936	53.21
Investment or exchange banks	126,616,122	1.40	81,496,018	2.30	116	0.52	0	0.00
Single credit cooperatives	205,546,932	2.27	109,736,208	3.10	0	0.00	5,214	46.74
Central and/or confederation of credit cooperatives	88,641,225	0.98	1,292,577	0.04	0	0.00	6	0.05
Development banks	814,809,767	9.01	318,581,606	9.01	13	0.06	0	0.00

Source Central Bank of Brazil. Available at https://www3.bcb.gov.br/ifdata/

that is only slightly lower than the points of commercial banks. This number gives an idea of how credit cooperatives are more diffused in the territory and, by working with a clientele in fact composed of individuals or smaller firms/agents, do not move relatively high values. But this proportion gives us the exact idea of how important they are to these local economic actors.

3.3 Community Banks: Finance to Empower Populations in Their Living Places

Brazil may be the country where community banks have had the most dynamic development in this twenty-first century. There are two main causes of this diffusion: (1) the incapacity of commercial banks to reach regions and places with a predominance of low-income population (and low economic dynamism) all over the national territory; and (2) the political mobilization of several communities to gain access to financial products and services, and thus be able to put into practice locally created collective projects to achieve this. Based on the pioneering experiences developed by Banco Palmas in Fortaleza (Ceará state)—as discussed in the second part of this item—more than a hundred communities in Brazilian cities organized and created their own financial institutions, with predominately local nature, establishing practices based in the "solidarity economy" (Singer 2002, 2009).

According to the definition of the two main scholars of the subject in Brazil, the Community Banks are institutions with "practice of solidarity finance to support the popular economies located in territories with low human development index," they are "structured from local associative dynamics" and aim to "generate and increase the income in the territory" (de França Filho and Silva Jr. 2009, 31). For the Instituto Palmas—the main institution that today organizes Banks of this type in Brazil—Community Banks can be defined as organizations that provide:

> Solidarity and networked financial services, of an associative and community nature, directed towards the generation of work and income in the perspective of reorganization of the local economies, based on the principles of a Solidarity Economy. Its objective is to promote the development of low-income territories by encouraging the creation of local production and consumption networks. It is based on supporting the initiatives of the popular and solidarity economy in its various spheres, such as: small productive enterprises, provision of services, marketing support, and the vast field of the small popular economies (Instituto Palmas, 2019).[7]

Neiva et al. (2013b, 113) point out that community banks are "institutions that result from popular and community organization and produce innovative arrangements, linking local culture and history to the production of the community's own development." Their actions help confer on financial institution other operational criteria, which broaden the impact of the Banks' action beyond its own economic

[7] Available at: http://www.institutobancopalmas.org/o-que-e-um-banco-comunitario/ (Accessed on 31.01.2019).

and financial dimension, since they also have a strong territorial, political, and cultural component. Due to these features, the action of these banks has an essential geographical logic. The "success" of the enterprise can only occur if the local population is engaged, and the production and consumption circuits of the neighborhood where the bank is located are dynamic. Santiago (2003, 31) is emphatic in recognizing the "exceptional" ethical attributes and territorial dimension as two of the central components of the development of community banks in Brazil (especially Banco Palmas). For the author, at least four specific activities developed by these banks are essential that can be important mechanisms for building a more just society:

1. The experience must have expressive popular leadership;
2. It aims at local development, based on identifying both consumption patterns and the productive possibilities of local actors;
3. Its central parameter is the "strengthening of the local economic space";
4. The notions of cooperation and self-respect of the population involved are central, and they become "social capital," which is also the basis of the dynamism achieved.

The first community bank created in Brazil was Banco Palmas, in the municipality of Fortaleza (state of Ceara), in January 1998. The recent spread of this type of institution, however, would only occur years later, and is directly related to the institutional support provided, from 2003, at the beginning of the government of President Luis Inacio Lula da Silva. In that year, the National Secretariat of Solidarity Economy (SENAES) was created, whose main function was to implement public policies that generate employment and income from practices linked to the "solidarity economy." The first, and only, Secretary of this federal agency was the economist Paul Singer, the main theorist of solidarity economy in Brazil (Singer 2002, 2009, 2013). Singer was the National Secretary of Solidarity Economy for 13 years, from the creation of the agency (2003) until its closure in 2016 (with the impeachment of former President Dilma Rousseff). According to the Legislation that the creation of SENAES is based on (Law nº 10683 of 05/28/2003), its main attributions were to

1. Support the definition and coordinate the solidarity economy policies within the scope of the Ministry of Social Development (to which SENAES was linked);
2. Articulate with representatives of civil society to help them determine guidelines and priorities for the policy of solidarity economy;
3. Plan, control, and evaluate the programs related to the solidarity economy;
4. Stimulate the creation, maintenance, and expansion of job opportunities and income generation and distribution programs, through self-managed enterprises, organized collectively and participatory, including activities already developed in terms of the "popular economy";
5. Contribute to microfinance policies, and stimulate credit cooperativism and other forms of organization for this sector.

The entire action of the Secretariat, therefore, was focused on the implementation of policies that were within the scope of the already mentioned "solidarity economy," including stimulating the creation and diffusion in the national territory of community banks, which fit perfectly in this profile of policies that aim to promote the empowerment of the poorest communities, as well as generate work and income for this same population. With the institutionalization of this federal government support in 2003, Banco Palmas also created the Palmas Institute, aimed at developing an operational methodology of solidarity finance, along with establishing partnerships and agreements with national and international institutions and NGOs that could assist Palmas' initiatives.

In this scale of national action, the Central Bank of Brazil itself changed its stance and regulation since 2003, with regard to implement initiatives to the financial inclusion of low-income population, including a more favorable view about the creation of community banks and local currencies in the country. The Bank started to organize discussion seminars on "microfinance" and began to have as one of its main concerns to make the banking system more inclusive (Feltrim et al. 2009). Between 2002 and 2013, the Central Bank organized 13 seminars to discuss and disseminate the idea of "financial citizenship," since commercial banks did not serve at that time a significant part of the population (Freire 2013, 43), as already mentioned. In 2009, the Central Bank started to establish partnerships with the National Secretariat of Solidarity Economy to improve the local currencies already existing in the territory (De Paula 2015, 66). In this way, instead of curbing popular initiatives linked to the solidarity economy, the Central Bank has become a "sympathetic" institution of these initiatives.

Another important event that explains the diffusion of community banks in Brazil was the partnership established between the Instituto Palmas and the Banco Popular do Brasil (Popular bank of Brazil) in 2005.[8] The Banco Popular do Brasil was created by the Federal Government at that time to promote financial inclusion as well. It did not have branches and operated through partnerships with other microfinance institutions (MFIs), as was the case with partnerships established with several Community Banks. With this combined action—and an initial portfolio of R$ 700,000.00 to be used with banks of this type—the number of community banks in Brazil increased to 37, with the following distribution in the territory: 25 in the state of Ceará (where is the headquarter of Banco Palmas), 4 in Espírito Santo, 3 in Piauí, 2 in Bahia, and 1 in the state of Mato Grosso (Neiva et al. 2013a, 10).

The great boom in the creation of community banks in Brazil occurred in 2005, based on a partnership between the National Secretariat for Solidarity Economy (SENAES) and the Palmas Institute, when the Brazilian Network of Community

[8]The Banco Popular do Brasil was a financial institution created in 2003 as a subsidiary of Banco do Brasil, exclusively aimed at promoting the banking and financing activities of low-income populations. The Federal Government created it in an effort to develop microfinance institutions focused on financing activities that generate work and income in the poorest cities/communities in Brazil. The federal institutions that focused on the promotion of microfinance were (1) National Economic and Social Development Bank—BNDES (with specific credit lines), (2) Banco Popular do Brasil, (3) development agencies, and 4. public banks with specialized portfolio, such as the Crediamigo program of Banco do Nordeste (Soares and Sobrinho 2008, 141).

Banks (Rede Brasileira de Bancos Comunitários) was created. The Network developed and "exported" the methodology created at Banco Palmas to several other places to create banks of this type in almost all regions of the national territory. In 2006, the number of community banks still increased and officially became known as Community Development Banks (CDB). Credit lines were established in public commercial banks to stimulate the opening of new units, mainly through two Public Notices promulgated by SENAES. The first Public Notice was SENAES/MTE no. 3 of 2010 and the second one the SENAES/MTE no. 1 of 2013. Both aimed to finance and create institutional conditions for new community banks to be opened, as well as to encourage the use of local currencies by these same banks (Neiva et al. 2013a). It is also important to highlight two other main partnerships with federal public banks that stimulated the development of community banks in Brazil in 2010 (Neiva et al. 2013b, 128/129):

1. The National Bank for Economic and Social Development (BNDES) opened a credit line of R$ 3,000,000.00 (about US$ 797,320.98) to the Palmas Institute to expand its operations through the Brazilian Network of Community Banks;
2. Caixa Econômica Federal (CEF), the largest public saving bank in Brazil, established a formal agreement with Banco Palmas, and a CEF banking correspondent was installed in the local branch, allowing the Banco Palmas to also provide "commercial" services, increasing the flow of people from the community into the branch.

This new degree of commitment of the Federal Government with propositions linked to the solidarity economy thus strengthened the initiatives of various social actors who already practiced actions with this content in different parts of the territory. According to De Paula (2015), most of the community banks in Brazil were created in this context, between the years of 2011 and 2014. Neiva et al. (2013a, p. 10) show that with these policies started in 2010, 60 existing community banks were able to consolidate their activities, and another 43 were created throughout the Brazilian territory (with the exception of the Southern Region of the country, best served by credit unions/cooperatives). The resources were used for three main functions: hiring of employees by the banks, contracting of technical assistance, and the purchase of infrastructure materials for the facilities.

Carolina de Paula's studies (2015, 2018) provide a detailed analysis of the distribution and functioning of Community Banks, indicating that there are nowadays 116 community banks in Brazil. Of the 27 states of the Brazilian federation, only 4 of them do not have any such kind of bank: Rondônia, Pernambuco, Paraná, and Santa Catarina. The Table 3.5 indicates that most banks are located in states in the north and northeast of the country, precisely the less developed regions and with the worst provision of services by the commercial banking system. The largest concentration of community banks is in Ceará, where 31% of the total (36 banks) are located; next came the states of Pará (10%), Espírito Santo (8%), Amazonas (7%), and Bahia (7%).

The Map 3.1 also gives a more precise idea of this distribution of community banks in Brazil.

Table 3.5 Brazil—Distribution of community banks by state (2019)

State	No. of banks	Banks in the state capital
Ceará	36	5 in Fortaleza
Pará	12	4 in Belém
Espírito Santo	10	2 in Vitória
Amazonas	9	6 in Manaus
Bahia	9	1 in Salvador
São Paulo	7	4 in São Paulo
Rio de Janeiro	6	1 in Rio de Janeiro
Distrito Federal	3	None in Brasília
Mato Grosso do Sul	3	None in the capital
Mato Grosso do Sul	3	None in the capital
Minas Gerais	3	None in the capital
Acre	2	2 in Rio Branco
Amapá	2	1 in Macapá
Paraíba	2	2 in João Pessoa
Piauí	2	None in the capital
Rio Grande do Sul	2	2 in Porto Alegre
Roraima	1	None in the capital
Rio Grande do Norte	1	None in the capital
Sergipe	1	None in the capital
Maranhão	1	None in the capital
Goiás	1	None in the capital
Total	116	30 in capitals

Source De Paula (2018) and Instituto Banco Palmas, 2019. Available at http://www.institutobancopalmas.org/rede-brasileira-de-bancos-comunitarios/

As in the case of Banco Palmas—that will be more detailed in this item—the strong associative and political action of the communities in question is present in the genesis of all other community banks in Brazil. The study by Menezes and Crocco (2009, 389) found, in the case of the two existing banks in the state of Espírito Santo—Banco Bem and Banco Terra—that the individuals involved were "residents of the neighborhoods that are on the margins of the economy and society" and struggled to improve local conditions in their neighborhoods.

De Paula (2015) also revealed that all four community banks in the city of São Paulo were created from local society and social movements initiatives, with a strong political claiming character. Three of these banks—Apuanã, Paulo Freire, and Autogestão (located in the North, East, and South zones of the city of São Paulo, respectively)—are directly linked to social movements for housing claims. The fourth existing Community Bank—Banco União Sampaio, located in the Southern Zone

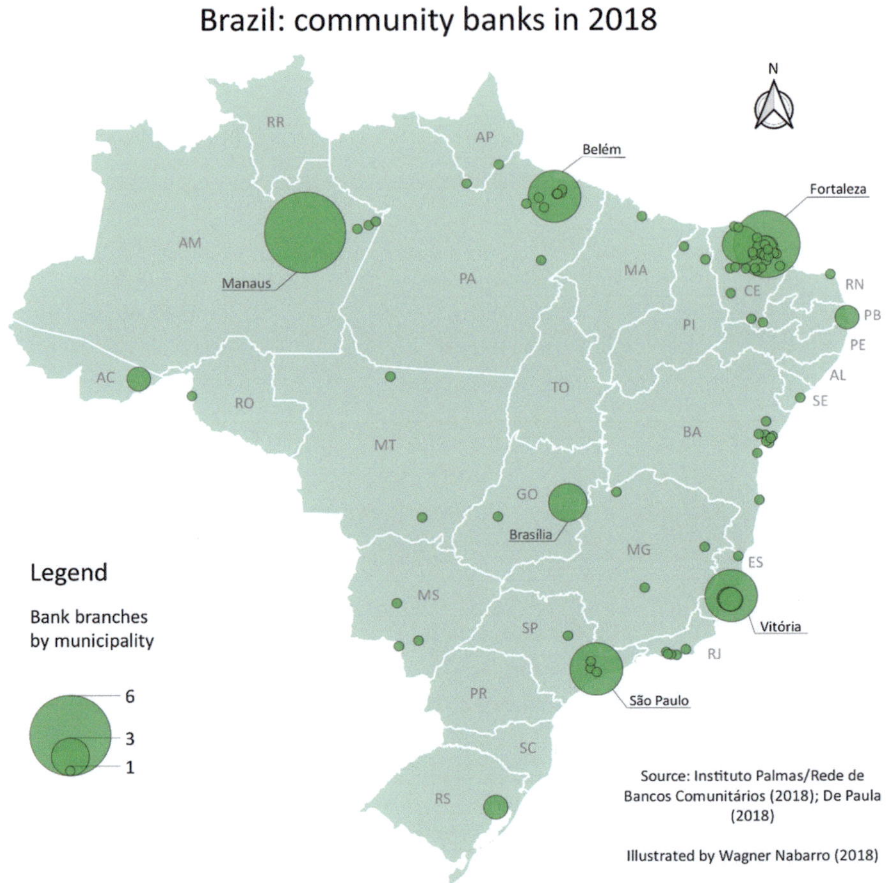

Map 3.1 Brazil—Community banks topology (2018)

of the municipality—originated from the action of the Popular Union of Women (UPM—*União Popular das Mulheres*), an organization officially founded in 1987 by women from the Bairro de Campo Limpo to demand from the public authorities to "pay attention to the serious problems that the region had such as worm diseases, malnutrition, children mortality, lack of day-care centers, health centers, schools, among other shortages of public services."[9]

Another important aspect to be highlighted is the participation of other public institutions—mainly public universities, both federal and state—in these networks

[9]In addition to being responsible for the creation of the Banco Comunitário União Sampaio, the Popular Union of Women "has a wide intersectoral activity that makes it present in several segments of the community. It develops a wide range of activities and initiatives aimed at the emancipation of women, equality in social, and gender relations, and the full realization of social, economic, political, environmental, and cultural rights." Available at https://uniaopopmulheres.org.br/quem-somos/ (Access on 19/01/2019).

that are on the basis of the implementation and continuity of the operation of Community Banks in Brazil. In the case of Banco Palmas, the activities of the Federal University of Ceará (UFC) were important several times, mainly for the development of management techniques and methodologies for the operational functioning and periodic evaluation of the Bank's results. The two other public universities that took actions directly linked to the development of community banks in Brazil were the Federal University of Bahia (UFBA)—mainly through the Technological Incubator of Solidarity Economy and Territorial Development Management (ITES/UFBA)—and the University of São Paulo (USP), through the Nucleus of Support for the Activities of Culture and Extension in Solidarity Economy (Núcleo de Apoio às Atividades de Cultura e Extensão em Economia Solidária, NESOL/USP).

A final point deserves attention when discussing the diffusion and recent operation of the community banks in Brazil. Several of them also use their own currencies, the so-called "local" or "social" money, which have the function of payment in local shops and services, empowering community banks, as well as conducting community consumption to occur within the neighborhood itself, to avoid the flow of locally generated surpluses being drained to other consumer circuits "outside" the bank's area of action. Local currencies began to be used in the 1980s in England and the United States, and in the 1990s in Argentina under different names and forms of use. In the middle of the first decade of the 1990s, there were more than 2,500 experiments of the type being carried out through Local Exchange Trading Systems, LETS (Tickel 2000). Since then, they have been given various names, such as "non-standard local monies" (Lee 1999), "alternative," "complementary," or "Community currencies" (North 2005, 2017). Local currencies are a kind of "agreement" in the community for the use of an exchange equivalent that coexists alongside the national currency of legal tender and serves as a local means of payment. The recent impetus of these practices is also related to the "rediscovery" by the post-Keynesians of the importance of money in economic dynamism (Chick 1994; Dow 1999; Amado 2006) as well as by the very recognition that "money is the most geographical of economic phenomena" (Menezes and Crocco 2009, 378).

The study by Rigo and França Filho (2017) provides a detailed analysis of the distribution and use of social currencies by community banks in Brazil. Seventeen community banks using social currencies were surveyed, 12 in the state of Ceará, 4 in Espírito Santo, 1 in the Federal District, and 1 in Bahia. One of the main findings of the study is that the difficulties are much greater than the "solutions" brought about by the use of local currencies. The two main difficulties are (1) to establish funding for the social currencies (constitution of a "credit fund") that allows it to be used in the community during longer periods of time and (2) to convince local sellers and shopkeepers to accept the social currency, thus allowing a more efficient circulation of it. For the authors, a greater acceptance would depend on the continued effort of banks to promote campaigns to raise awareness of both consumers and traders so that the currency could effectively present liquidity and be increasingly accepted in the community. The results, therefore, are not encouraging, even in the case of Banco Palmas, the most developed of the Brazilian community banks.

Because of these difficulties, the community banks themselves are gradually replacing the bills of social currencies for digital ones, known mainly as e-money (e-dinheiro). At the very basis of this use of e-money was the creation in 2014 of the Instituto Banco da Periferia (IBP) by the founders of Banco Palmas. As shown by the Institute's website, the IBP it is a Non-Governmental Organization (NGO) founded in the framework of Banco Palmas, mainly aimed to develop and disseminate a digital platform that allows the use of this digital currency by other community banks. The platform provides the condition to manipulate/manage e-money with a smartphone, or even a cell phone that does not have access to Internet. According to the definition by the Palmas Institute, e-money "is an electronic social currency, operationalized using a smartphone, preferably, to enable, through the installation of an application specially developed for this purpose," and is nothing more than "a new means of digital payment."[10] As the system's slogan shows, e-money is "the community bank in the palm of your hand!"

With this information system, it is possible to carry out virtually all operations and have access to all the services that a commercial bank offers. The main functions available to the user are deposits, bills payment, money transfers, and account statement (Instituto Palmas 2019). Moving the paper-based social currency and/or written records or documents to a digital platform is one of the main hopes of community bank managers to modernize their systems, consolidate their financial sustainability, and increase the communities' confidence in the use of local digital currencies. In the 20-year manifesto of Banco Palmas, the high expectation regarding the use of e-money points to the possibility of widespread diffusion of this financial innovation, for other initiatives of the same kind throughout the territory:

> The digital strategy allows us to think on a larger scale aiming at the sustainability and financial independence of the Community Banks. We have entered the world of virtual currencies, cryptocurrencies, Blockchain, and other models. We believe from our own experience in decentralized and distributed systems, especially when we speak of the financial world, and of course in the democratization of money production, which today is held by a private and non-transparent system (Banco Palmas 2018, 6).

As shown by De Paula's study (2018), currently in Brazil 27 community banks use e-money, the main ones being Banco Palmas itself, but also Banco Tupinambá (Belém-PA), Banco União Sampaio (São Paulo-SP), Paulo Freire (São Paulo-SP), Autogestão (São Paulo-SP), and Mumbuca (Maricá-RJ). In 2016, a total of around R$ 10,000,000.00 in e-money were exchanged in these banks, reaching an average of 2,138 users of this digital social currency. In addition to Banco Palmas, most successful experience is the Banco Mumbuca located in the city of Maricá, in the Metropolitan Region of Rio de Janeiro (103 km north of Rio de Janeiro).

It is not just about finance: local social mobilization in the implementation of Banco Palmas

As was stressed at the beginning of this item, Banco Palmas is the largest community bank in Brazil, having become a complex organization, which currently "exports"

[10] Available at: http://www.institutobancopalmas.org/wp-content/uploads/moedas-TRANSI%C3%87%C3%83O-vers%C3%A3o-final.pdf. Accessed on 19/01/2019.

its expertise and information/management systems to other banks of the same type all over the Brazilian territory. This robust action of this Bank is responsible for the large number of other community banks in the state of Ceará as well. Banco Palmas was created on January 20, 1998, in the Conjunto Palmeiras neighborhood, a poorer suburb of the city of Fortaleza, in the state of Ceará. Conjunto Palmeiras, in its turn, was created in 1973 and was the result of a large removal of families who originally lived in Favela Lagamar, located in the center of the city of Fortaleza, on the banks of the Cocó River and Avenida Beira Mar (an extremely valuable location from the point of view of real estate sector). Altogether 1,500 families, who lived in this central area of the city, were removed to the Conjunto Palmeiras in 1973. At the beginning of its creation—and due to its peripheral location—, the new neighborhood had few public facilities and infrastructures, with precarious conditions of transport equipment, lighting, sewer, and other elements of the local built environment (Toscano 2003). In this context of enormous shortages, in 1977 and 1978 the first forms of organization and mobilization of the local inhabitants appeared, aimed at improving the conditions of the built environment of the neighborhood, and demanding more public services in the area.

These population struggles helped create the Association of Residents of Conjunto Palmeira, (ASMOCONP—Associação dos Moradores do Conjunto Palmeira), on February 2, 1981. With the creation of the Association, "the popular mobilization was amplified and the struggles for various services were generalized" (Toscano 2003, 11). The ASMOCONP was responsible for the creation of Banco Palmas. The first source of funds to create the Bank came from a Ceará State NGO—based in Fortaleza—*Cearah Periferia*. Through a "Fund for Support to Self-Management Projects—FAPAG," this NGO provided the recourses to ASMOCONP that gave rise to Banco Palmas. At the beginning, R$ 2,000.00 (currently about US$ 530) was made available for the creation of the Bank, to be paid in 1 year, with moderate interest of 1% per year (Segundo and Magalhães 2003, 16).

In the context of the creation of Banco Palmas, other activities and micro-institutions were organized as well, to raise awareness and mobilize the population to confront the enormous problems that the Conjunto Palmeiras faced. Most of these actions aimed to influence public authorities in order from them to improve local conditions of living into the neighborhood. Initiatives such as the *Memory of Our Struggles Program*, *Prorenda Project*, *Inhabiting the Uninhabitable* Seminar, *Integrated Community Development Plan* (PDCI), and the creation of a local community newspaper (the "Desperta Palmeiras") were implemented in this period. In the second *Inhabiting the Uninhabitable* Seminar, organized by the community in 1997 (the first seminar was held in 1991), one of the conclusions drawn was that "the neighborhood was already urbanized but the economic poverty of the residents had increased" (Segundo and Magalhães 2003, 14). The second important "conclusion" was the necessity to create projects that generate work and income for the local population, which explains the creation of Banco Palmas. In the same period, the community leaders participated in the First Brazilian Meeting of Culture and Socioeconomics Solidarity, held in June 2000 in the city of Mendes, Rio de Janeiro state. This meeting was the cornerstone of the creation of the Brazilian Network of Socio-Solidarity Economy, which Banco Palmas became part in the same year.

With the incorporation of the principles of solidarity economy in the creation of Banco Palmas, all the actions carried out were concerned with social and economic development of the community, starting from a more integral vision of the process, combining the instrumental logic of finances with the necessities of the community at large. The first problem that the Bank aimed "to solve" was the large number of people who lived in the neighborhood in productive age without occupation. It was found that most of these individuals, despite not having complete formal education, had the skills to develop various economic activities, such as handicrafts, apparel, cooking, construction, personal and domestic services, among many other "jobs". Another major problem was the fact that the few existing producers of Conjunto Palmeira could not sell their products in the neighborhood, given the low income of the residents, as well as the fact that the residents did their shopping in other neighborhoods (where they found cheaper prices). Thus, a greater problem than conditions of local agents to produce was the lack of channels to sell their production. To organize and to "channel" local consumption within the community in order to reinforce local demand (and thus create markets for local production) was also seen as a way of keeping the existing wealth locally, avoiding its drainage to other neighborhoods/economic activities of the city.

To bring these two perspectives together—stimulating both consumption and local production—the first financial innovation of Banco Palmas was created: the Palmacard. The Palmacard was a paper-based credit card containing basic information about the owner of the card (on the front) and the notes on the products purchased and their values (on the back of the card). The use of Palmacard was restricted to local establishments that had previously registered with Banco Palmas and accepted payment with the card for products/services offered. At first, 20 families received the Palmacard, and 5 producers with credit from the Bank created. The cards had initial value of R$ 20.00 (around US$ 5) and could reach up to R$ 100.00 (if the user was punctual in their payments). No fees were charged for the use of the card, nor were interest charged on the small loans made (Segundo and Magalhães 2003, 36).

From the point of view of the Bank's funding, the first major injection of resources occurred at the end of 1998. After the success of the initial actions of Banco Palmas in the community, the first "international partners" began injecting funds into the small institution. These were Oxfam (United Kingdom) and GTZ (Germany). This contribution allowed the bank to end the year 1998 with approximately R$ 15,000.00 in its portfolio (US$ 3,980) and 120 credit cards operating in the hands of resident borrowers. Thus, the entire genesis of the activities of Banco Palmas had a strong relationship with the organization of the local population, but also with the contribution of resources from international solidarity organizations.

At the time of its creation, the Bank's actions were publicized in Conjunto Palmeira through the neighborhood radio and member assemblies, to mobilize residents to engage in activities, and thus create a "local network of solidarity economy" (Segundo and Magalhães 2003, 18). Concomitantly with this continuous effort of communication and mobilization of the residents, the Bank carried out a systematic survey/mapping of the main forms of production and the products typically consumed by the families living there. This survey was called the "mapping of production and consumption" of Conjunto Palmeira. The numbers collected by the survey were "surprising" in relation to the amounts involved. For the year 2010, the survey verified that the amount of R$ 1,540,251.88 (which corresponded to approximately US$ 440,000) was spent monthly by families on commerce in the neighborhood. Another fundamental feature of this amount spent monthly was that R$ 1,015,000.00 (or about 66%) was spent on food. The most logical step was to develop in the community a system of incentives for local production of these alimentation needs. The credit system then operated initially on two fronts: (1) a microcredit system aimed at financing small local productive activities (mainly food products) and (2) consumer credit instruments (mainly the Palmacard).

In addition to the system for small loans, another actions were also implemented in the Conjunto Palmeira to enable these new conditions for the circulation of a local currency and goods, such as the creation of fairs featuring local producers, a solidarity store (which operates in the same Banco Palmas building), and other "awareness" actions to convince the population to do their shopping in the neighborhood. Finally, part of this "holistic" approach of Banco Palmas, from the beginning of its activities, included providing for the population in the neighborhood teaching and professional training programs, a technological incubator (for women at vulnerable situations), a Laboratory of Urban Agriculture, and a barter club to facilitate the use of the Palmas social currency (Segundo and Magalhães 2003, 18).

Another interesting feature for the functioning of the Bank concerns the eligibility criteria for access to credits granted. These criteria sought to break from the typical instrumentality of commercial bank evaluations, incorporating elements related to the neighborhood life and the geographical proximity of the actors. Among the main criteria, the following stand out:

1. Being a resident of Conjunto Palmeira and a member of the Association of Residents (ASMOCONP), a simple operation to be carried out with the completion of the registration form. The only obligation required by the Association was participation in Assemblies and engagement in community activities.
2. Be a "recognized person" in the community, that is, have the approval of your neighbors. The Bank therefore does not use official consultation instruments from established "credit protection" agencies (or personal credit rating companies), such as Credit Counseling Service (SERASA), Credit Protection Service (SPC), Informative Register of Unpaid Credits (Cadastro Informativo de Créditos Não-Quitados, CADIN), among others, since most of the population of the neighborhood has restrictions and are "blacklisted" in these agencies. The Palmas Bank, therefore, does not require a "guarantor," as is often the case in commercial banks. The main technique of credit analysis is the consultation with the neighbors of the interested party, "who provide the guarantee about the credibility of the person" (Segundo and Magalhães 2003, 35).

3. To sign a "social contract" with the Bank, which commits the client, whenever possible, to buy and sell their goods within the limits of the neighborhood (Segundo and Magalhães 2003, 34).

In addition to the aforementioned Palmacard, the Bank initially offered four other main lines of credit:

1. Microcredit for production, trade, or services: credit aimed at the interested party who wants to start/create some productive activity in the neighborhood and who has not been able to obtain credit from traditional commercial banks (for failing to meet the formal eligibility criteria that these institutions require); the limit of the value of this type of credit is R$ 1000.00 (about US$ 265).
2. Microcredit for women: a specific line aimed at women whose families are in extreme deprivation/poverty situation. The credit is released upon request made by the interested party(s) in the Women's Incubator of ASMOCONP.
3. PalmaCasa: credit line for small housing reforms, released after presentation of the list of materials to be purchased. The borrower has 6 months to repay the loan, with interest of 1.5% per month.
4. Urban agriculture: credit line for agricultural projects carried out in the neighborhood, mainly in the dwellings of the residents (in their backyards); both the cultivation of legumes and vegetables, as well as medicinal plants or chickens for meat and eggs are encouraged (Segundo and Magalhães 2003, 37).

According to a survey carried out recently, in celebration of the 15 years of operation of Banco Palmas (Neiva et al. 2013b, 141), the main financial services currently offered by the bank are as follows:

1. Correspondent Banking and ATM of Caixa Econômica Federal and Banco do Brasil (the two largest Brazilian public banks);
2. Palmas microinsurance;
3. Electronic social currency;
4. Productive credit;
5. Consumer credit in social currency.

As mentioned, the Palmas bills were replaced by Palmas e-money, an all-digital "electronic social currency" that can be managed by smartphones and aims to make local exchanges more efficient and versatile. The literature analyzed, as well as the descriptions made by the agents themselves, show that the introduction of these information techniques at the very heart of the bank installations and procedures can make the initiatives even more significant from the point of view of their results for the full development of the local community. According to the Palmas Institute itself:

> Palmas e-money is an electronic community bank, without branches, with unlimited reach. Numerous developments in the use of the platform are planned, to be deployed after its launch. They are: Payment in the local commercial network; Payment of public services, such as public transport; Bill payment; Payment of social benefits by municipal, state, and federal governments. With the mass use of the system, a channel of instant communication

with the community is created, which allows projects related to health and financial education to be added (Instituto Palmas 2019)[11]

Banco Palmas is, therefore, a concrete example of how the financialization of the territory can take place on other bases, not totally linked to the corporate and capitalist instrumentality that is typical of regular commercial financial institutions. Their action goes beyond the strict financial dimension and has broader and ethically interesting repercussions, such as empowering poor communities, increasing their self-respect, valuing local production and consumption circuits, concern for the women and the economically more vulnerable, as well as promoting constant campaigns of professional training and political awareness. The same reasoning applies to all other community banks that exist today in Brazil.

3.4 Local Action and Global Tools: The Fintechs

When analyzing the consequences of massive introduction of information techniques in the productive processes, Lojkine (1992) and Hepworth (1990, 27) shows that there is also a new logic of location of the economic activities' facilities and of the workforce itself, due to the changes provided by the diffusion of computers and of information transmission networks (that significantly altered the previously existing standards). In the so-called "banking industry," this disruptive power of information has also been very intense, and the fintechs are the most thorough form of this phenomena. But what are fintechs after all?

Fintechs are not an easy phenomenon to be defined. They don't even appear in the former dictionaries of economy and finance, and the term also does not exist even in the 2008 edition of *Dictionary of Finance and Banking* from Oxford (Law 2008). Under the etymologic point of view, fintech is simply an abbreviation that joins the English words "financial" and "technologies." The neologism appears in the contemporary context of banalization of information techniques, which profoundly changes not only the financial sector, but practically all economic activities, from agricultural to industrial and trade and services. All companies of high technological content that recently emerged to provide services of financial nature, whether in a complementary form to the "traditional" financial institutions, or to compete with them, may be considered as fintechs. For Tsai and Peng (2016, 7), the fintechs may be considered simply as a result from "combination of financial services with information technology." Although most of these companies are relatively small, they have enormous potential to completely change the contemporary financial business model (Hendrikse et al. 2018).

According to the Brazilian Association of Fintechs (ABFintechs), fintechs may be defined as "companies that use technology intensively to offer products in the area of financial services innovatively, always focused on experience and user' needs."[12]

[11] Available at http://www.institutobancopalmas.org/palmas-e-dinheiro/ (Access in 23.05.2018).

[12] Available at https://www.abfintechs.com.br/sobre.

The Central Bank of Brazil (BCB 2017, 99), on its turn, defines fintechs as "*startups* intensive in the use of financial technology." For the Financial Stability Board (FSB), an entity directly linked to Bank for International Settlements (BIS), fintechs may be defined as "Financial innovations, enabled by technologies that may result in new business models, apps, processes or products with tangible effects on markets, on financial institutions and on the provision of financial services" (BCB 2017, 99).

The fintechs are therefore, in first place, a new economic actor typical of the informational period itself—or technical-scientific-informational period, as demonstrated Milton Santos (1994a, b)—in which we are living. With the massive diffusion of information techniques and all objects and technical systems that better represent it (such as data centers, satellites, fiber optics, Internet, computers, laptops, smartphones, etc.), both the costs of their use decrease, as well as increases the so called "learning curve" of the population regarding its use. These conjugated processes enhance even more the diffusion of economic activities and firms such as fintechs. As recently showed by Haddad and Harnouf (2018), the basis for the diffusion of fintechs (and of startups in general) is linked to new "fixed" infrastructures and "movable" technical objects:

1. Circulation infrastructures and information storage: "Access to supporting infrastructure such as broadband networks might be of crucial importance for the emergence of fintech in a country" (Haddad and Harnouf 2018, 3/4);
2. Technical objects systems that serve as interface of companies with the consumers and users (first personal computers, afterward mobile phones, currently tablets and smartphones). For the authors, "almost inconceivable growth in mobile and smartphone usage is placing digital services in the hands of consumers who previously could not be reached, delivering richer, value added experiences across the globe" (Haddad and Harnouf 2018, 4).

On the basis of the growth of these new activities is also the managerial change caused by the entry in the consumer market of the so-called *millennials*, i.e., the new generation of individuals born as from the 1980 decades, and that already arrived into the world surrounded by this technical-scientific-informational *milieu*. This generation has certain types of social, cultural, and economic behavior pattern that are greatly influenced by this new kind of environment. Researches made by the banks themselves show that these consumers simply mention that "will not need" banking services in short and medium term, and that the products and services offered by the traditional banks do not meet their needs, as well as the channels and the communication forms of these banks do not speak the "language" of millennials (VALOR ECONÔMICO 2018). In an article about the incorporation of information techniques in "traditional" banks, Accorsi (2014, 204) shows that there is a profound alteration of the consumer profile of financial services, directly connected to the changes in behavior of "generation Y and Z," which are already obligating the banks to revolutionize their forms of distribution and use of banking services, with strong utilization of other interfaces/communication channels (that incorporate attributes of social networks) and that also make available new types of more "customized" products and services for this younger generation.

These changes reflect in practically all sectors of the economy, mainly in the countries in the core of the capitalist system. To mention only a few examples, the streaming companies, such as Netflix (leisure/cultural industry), the booking system of Airbnb (tourism industry), Uber (urban transportation), and Amazon (e-commerce), are examples of large companies which started as local small startups. This actual invasion of computer systems and apps has been considered in the information economy as the fourth industrial revolution, in which automation will gain even greater proportions, and the knowledge and information have become the main commodity for economic circuits.

With regard specifically to fintechs, under the conjunctural point of view, another essential factor for its emergence was the financial crisis of 2007/08, when "traditional" operators of financial system, mainly big commercial and investment banks, experienced huge economic losses, and saw their legitimacy and trustworthiness demolished within common sense/public opinion. This crisis paved a way for the emergence of other service provider agents, with the purpose of offering alternative forms of treating finances and use of financial services. For Wójcik and Cojoianu (2018, 218), "The crisis created conditions for the rise of fintech as it made the established financial sector more vulnerable to competition from technology firms and new start-ups. The crises damaged the reputation of financial firms, particularly banks." The crisis was also a fundamental element for the re-adequacy of part of the financial labor force, wherein significant part of the workers in "traditional" financial institutions are exactly those that since then are creating their own companies, with smaller dimension, aiming to develop new processes or products for the proper financial industry (Fintechlab 2017, 5). Haddad and Harnouf (2018, 2) state also that there is a "large number of investment bankers who lost their jobs after the financial crises and are now eager to use their finance skills in a related and promising financial sector."

In Brazil—as in Latin America—the emergence of fintechs is a very recent phenomenon. According to the Bank of Interamerican Development (BID 2017, 17), "it is estimated that 60% of all young fintech companies that operate currently in Latin America emerged from 2014 to 2016," as also shown in Table 3.6.

For the analysts from Fintechlab (2017), the year 2016 was the "turning point" for the development of fintechs in Brazil, due to the number of companies that emerged in that year, as well as the increase of complexity of the division of labor with the fintechs. In the report on what was called the "Brazilian ecosystem of fintech," there are five types of main operators: the fintechs itself, the regulators, the associations of the sector, the investors, and the accelerators (Fintechlab 2017, 5). This division of labor, in general lines, may be defined and exemplified as follows:

Table 3.6 Latin America—Emergence year of fintechs (in 1000 units)

Start	Before 2011	2011	2012	2013	2014	2015	2016
%	10.9	7.0	7.7	14.4	15.1	23.2	21.8

Source (BID 2017, 16)

1. The public regulators are basically three: Central Bank of Brazil, Securities and Exchange Commission (Comissão de Valores Mobiliários, CVM), and Private Insurance Superintendence of Brazil (SUSEP);
2. The associations of the sector are mainly as follows:

   a. Brazilian Association of Fintechs (ABFintechs), created on October 25, 2016;
   b. Brazilian Association of Payment Institutions (ABIPAG);
   c. Brazilian Association of Digital Credit (ABCD), created in November 2016, as from the merger between credit fintechs Biva, Geru, Lendico, Creditas, Simplic, Trigg, and Just Bank (Fintechlab 2017, 14);
   d. Association of Equity Crowdfunding, known also simply as Equity, was founded by the companies Broota, Economia Criativa, EqSeed, Ideias de Futuro, Kickante, Start me up, The Pie, Urbe, and Viking, all destined to foment the practice of crowdfunding in Brazil (Fintechlab 2017, 15).

The Central Bank of Brazil, in their report and documents about fintechs, has had a positive and stimulating evaluation of the development of this type of small firms, for believing that the entry of new agents in the financial system may enhance its efficiency and reduce costs for the economy as a whole. The two main concerns of Central Bank—as well as of all other monetary authorities in the world—are the maintenance of the stability/solidity of institutions and financial inclusion. Due to both its technological disruptive potential and for the fact of being companies that mostly deal with financial assets, the close monitoring by the Brazilian monetary authorities is a central element for a positive coordination of changes that fintechs are bringing to the national financial system.

This incentive attitude for the insertion of intensive-information companies in the division of labor in financial sector is consigned in the recent bank report. According to the Central Bank, the action of the regulator shall not be "inhibitor neither permissive" (BCB 2017, 102), and in their assessment until the moment the current signs about the maturity of fintechs were ambiguous: there are those that argue that the market valorization of startups creates return expectations beyond justified by fundamentals, as well as consulting firms report an acceleration in direct investment and in acquisitions of startups by banks (BCB 2017, 102). As the Central Bank document shows,

> The Central Bank of Brazil (BCB) recognizes the importance of both employment of new technologies, which application may extend over the industry, including payment methods, compensation and liquidation, as well as the importance of innovative forms of service provision. Likewise, encourages the development of these new technologies in financial market, once this may stimulate the competition in market, what has impact in its efficiency and allows to offer products at lower prices to the clients, reaching greater part of the population (BCB 2016, 51)

On the other hand, the document demonstrates that the bank has similarly a "vigilant" attitude in respect to the introduction of these technical innovations and of business models "as far as they may have consequences over the soundness of the financial system" (BCB 2016, 51). Likewise, in a concrete point of view, the Central Bank has adopted the following measures:

1. Participation in international forums of financial institutions, which have been analyzing closely the impacts of fintechs in the world, such as mainly the case of banking supervision forums of Bank for International Settlements (BIS), through Basel Committee on Banking Supervision (BCBS), but also in the work group about digital innovations formed by CPMI and by the International Organization of Commissions of Securities (IOSCO), work group on financial technologies, in the ambit of Financial Stability Council (FSB), and Union of International Telecommunications (UIT) (BCB 2016, 51).
2. Promulgation of Ordinance BCB No. 89.399, from June 3, 2016, which created in the Bank an interdepartmental working group with the purpose of "elaborating studies about digital technological innovations" that interfere directly in the National Financial System and/or in the Brazilian Payment System (SPB) itself, evaluating the "potential impacts over the operation of these systems."
3. Creation of the Laboratory of Financial and Technological Innovation (Lift) on May 9, 2018. The referred Laboratory "is a collaborative virtual environment with the Academy, the market, the technology companies and the *startups*, destined for the development of technological novelties, to exchange knowledge and evaluate results of experiments" (BCB 2017, 101).

For being companies that provide financial services, but are not banking institutions, they are also a fundamental element of the recent process of "financial disintermediation," process that tends to directly undermine the action of commercial banks. The fintechs, therefore, are companies extremely innovative that provide simpler and easier services to be used by the general public. They have operational structures extremely "lean" and automatized, and, therefore, small relative size. It is this exact characteristic of high organic composition of the capital that allows the consecution of another of their main characteristics: its operational costs are low (what generated elevated internal economies) and are also lower the prices of products and services offered.

Another essential aspect of the fintechs is that they, in general terms, empower the client, especially for being small-size companies, which results in smaller differences between their economic size and that from the size of their regular customers. The opposite serves to understand the relationship clients/traditional commercial Banks: it is immeasurably higher the economic power of major banking institutions when compared to their individual clients (mainly small firms and individual customers).

The "differentiated" options offered by fintechs, in addition to being cheaper for the final customer, are less bureaucratic (once automatized), they usually adopt a more direct non-presence communication (through customized apps and software) and are much more transparent in terms of information disclosure that are "behind" the contracts (what also reduces the asymmetry of information between agents) (VALOR ECONÔMICO 2018). According to Fintechlab study (2017, 6), the massive introduction of information technologies allows the following improvements in the financial service provision system of fintechs:

1. Reduction of operational costs;
2. Greater "scalability" of arrangements;

3. Practice of lower prices that, in theory, also allow the financial inclusion of a larger share of lower income population.

The practice of lower prices in products and services is also seen in other sectors of activity of fintechs, such as, for example,

1. New payment systems, mainly with the availability of POS terminals, without charging fees/monthly payments;
2. Free check account services (or with lower membership fees);
3. Free credit and debit cards;
4. Personal loans;
5. "Friendly" digital collecting systems;
6. Fortune management services;
7. Insurances (VALOR ECONÔMICO 2018).

This new operation standard and the prices practiced are a threat to all financial agents considered as "traditional," especially commercial/retail banks. It is not possible—or is very difficult for them—to compete with fintechs in products and services that they start to offer, and the transformations in the financial market are not greater because of the official financial regulation of the countries, not allowing a more accentuated rupture caused by fintechs (Philippon 2017). For the traditional commercial banks, weighs the enormous inertia of their huge technical infrastructures (headquarters, operation units, branches, internal networks, contracted labor force), and a more bureaucratic internal division of labor, what becomes impossible to make them leaner to face the new organization forms of fintechs. According to the analysis of Interamerican Development Bank, this digital transformation with the fintechs is a clear threat to the financial sector in two main directions: (1) to the volume and profit margin of traditional operators, and (2) to the very existence of traditional operators (BID 2017, 6). According to the document, "it is witnessed the birth of a new wave of technological companies that compete in equal conditions with the traditional actors of financial services, such as the case of banks, insurance companies and payment networks" (BID 2017, 6).

Philippon (2017, 2) summarizes as follows the disruptive potential of the fintechs:

> Such innovations can disrupt existing industry structures and blur industry boundaries, facilitate strategic disintermediation, revolutionize how existing firms create and deliver products and services, provide new gateways for entrepreneurship, democratize access to financial services, but also create significant privacy, regulatory and law enforcement challenges.

Another particularity of fintechs is the form in which they are financed. Like the startups of other economic sectors, those that already stand out in the beginning of their operations (for presenting an actually innovator "business model" and growing potential) receive financing from investment funds of technology, whether of angel investors or large financial institutions (VALOR ECONÔMICO 2018). In almost the totality of the cases, fintechs that stand out with the proposal and execution of services and innovator products with major market potential are almost automatically bought by major banks and financial agents (undermining their effectively disruptive potential).

A calculation of FintechLab company [13] made in June 2019 shows that there was in Brazil in this year 529 fintechs, divided into the following operation niches:

1. Payment systems: 151 (29%);
2. Loans: 95 (18%);
3. Financial management: 90 (17%);
4. Investments: 38 (7%);
5. Insurances: 37 (7%);
6. Cryptocurrencies and DTLs (distributed ledger technology): 36 (7%);
7. Funding: 25 (5%);
8. Debt negotiation: 19 (4%);
9. Businesses with foreign exchange/currency and remittances:14 (3%);
10. "Multiservice" companies: 12 (2%);
11. Digital Banks: 12 (2%).

An example of a successful fintech recently created in Brazil was BLU365 (whose initial name was Kitado). The company was created in 2014 by a former director of Banco Santander in Brazil (Paulo de Tarso) and has as objective "to improve and humanize the credit recovery negotiations." The main "failure" identified by the manager, and that justified opening the firm, was the nature of the treatment provided to the clients by commercial banks. In these institutions, after 60 days of no commitment payment (a period that will characterize the client as being considered as defaulting in Brazil), the debtor client moves from the account manager's care to the collection department of the financial institution, starting of being pressured "by poorly prepared attendants that use an aggressive approach to coerce to make a negotiation" (VALOR ECONÔMICO 2018).

Another example is of the company Magnetis, which offers "automated investment services," with use of "advisor robots," also a new financial technology which has grown significantly in the past few years. Founded in 2015 is considered the first investment fintech in Brazil.[14] Companies such as Magnetis operate in the frontier of traditional financial segments such as Private Banking, wealth management funds, or "family office." The initial investment requirement is also relatively lower—starting at R$ 1,000.00—what increases the basis of possible clients (most of the clients invest from R$ 25,000 to R$ 400,000). A last advantage characterizes this type of fintech: once they are not offering their own products (they only manage the investments), they do not have conflict of interest in the investment recommendations that they make (contrary to large banks and managers, that many times "oblige" the client to purchase products from the bank itself, even when these products are not very profitable for the client) (VALOR ECONÔMICO 2018).

Another agent that appeared recently in this context of the informatization of the financial economyis the digital banks. Also according to Report FintechLab 2017, the digital banks are, in first place, the "response of traditional banks to the

---

[13] Available at http://fintechlab.com.br/index.php/2018/08/13/novo-radar-fintechlab-mapeia-mais-de-400-iniciativas/ (Access in 10.01.2019).

[14] Available at https://magnetis.com.br/quem-somos (Access in 20.03.2019).

entry of technology giants in the financial system." The mains specificities of the digital banks are as follows:

1. They are financial companies extremely intensive in information, and therefore have leaner infrastructures and lower labor requirements (mostly when compared to the traditional commercial banks);
2. The main interface (or channel) of service provision is the smartphone (hardware), which makes feasible the use of apps (software) developed by the banks;
3. Enable the "total management" of check accounts by the clients (in traditional banks, this management is shared with the account managers);
4. For acting basically through apps/software, they are extremely fast and flexible under the operational point of view;
5. The headquarters generally are installed in buildings with innovative internal layout, multifunctional rooms, and that facilitate communication among the different employees/bank sectors;
6. The funding has two main sources: one parent bank (when is part of a larger conglomerate) and major institutional investors specialized in startups (angel funds/investors).

Currently in Brazil, according to Fintechlab report of 2019, there are twelve digital banks in operation: BTG Pactual Digital, Banco Digital Maré, Banco Original, Neon, Banco Inter, Next, Sofisa Direto, BanQI, Modalmais, C6Bank, AgiBank and Nubank. Although digital banks are similar due to all technical characteristics and service provisions mentioned above, the ownership structure of each one is different. Banco Original, for example, is headquartered in São Paulo and belongs to JBS Group, Brazilian giant from processed meat segment. According to the bank site, "the culture and innovation essence are from a startup, but the experience is of a solid bank, with over R$ 8.8 billion in assets" (Banco Original 2018).[15] The bank was created with a "100% digital" structure, what allows greater agility in assistance and less costs in provision of services.

Banco Original emerges actually from the proper financialization of their parent company, JBS. In 2008, the company creates the Banco JBS, aiming to operate as a financial branch of the company. In 2011, JBS buys Banco Matone (a small bank headquartered in Porto Alegre), creating Banco Original. On the year 2015, the bank also directs its operation to retail (until then was specialized in assisting clients from agricultural and corporate value chains), launching Banco Original as the first 100% digital bank in Brazil. In December 2016, the bank had 232,000 account holders, in December 2017 the number was 542,000, and nowadays (June 2019) they have about 1.6 million customers.[16]

Banco Next was created in 2017, as an initiative of one of the largest Brazilian commercial banks, Bradesco. Next offers services and financial products like other

[15] Available at https://www.original.com.br/sobreobanco/ (Access in 21.01.2019).

[16] Avaiable at https://www.original.com.br/docs/relations/results/conglomeradoFinanceiroOriginal/2019/DF-Conglomerado-Financeiro-Original-1-semestre-2019.pdf.

digital banks, such as check account, savings account, credit and debit card, loans, investment options, and other "innovative" resources, possible to being performed as from the use of bank apps downloaded into smartphones (financial planning, discounts when shopping, etc.). The great "advantage" offered by the bank is exactly the fact of being allowed to use any branch of Banco Bradesco in national territory (currently 4,765 added to 1,440 service stations).

Once made these systematic approximations about fintechs in Brazil, what may be said in respect to their localization patterns? What are the main factors that influence the distribution and operation of fintechs?

For Haddad and Harnouf (2018), the economic development degree of the countries and the availability of modern technical infrastructures are the two main factors that explain the emergence and development of fintechs in the world. For them, the other factors that have positive impact in the creation of fintechs are as follows:

1. A well-developed economy with venture capital readily available;
2. Secure Internet servers;
3. Efficient system of mobile telephone subscriptions;
4. Available skilled labor force (Haddad and Harnouf 2018, 1).

Due to this dependence of economic and technical factors, under the point of view of their preferential localization, fintechs are currently concentrated mainly in the more developed countries, and also in the largest metropolis of these countries. The study made by Haddad and Harnouf (2018) also provides us a more precise dimension of the distribution of fintechs in the world recently (Table 3.7).

The report of Fintechlab (2017, 6) shows that the main financial centers where the fintechs are developed are, in this order, London, New York, Silicon Valley, Tel Aviv, Singapore, and Shanghai. In the Brazilian case, the most recent data provided by ABFintechs (2018) show the following regional distribution: São Paulo state locates the vast majority of fintechs (58%), followed by the states of Minas Gerais (9%), Rio de Janeiro (8%), Paraná (7%), Santa Catarina (6%), and Rio Grande do Sul (4%). All 21 other states of the federation concentrate only 7% of the country's fintechs, which reveals the enormous spatial selectivity of the location of this type of firm. According to other study from Fintechlab (2017, 11), São Paulo stands out mainly for offering "the main channels to capture resources, such as investment funds, venture capital funds, angel funds and the proximity with banks and access to entrepreneurs."

A final characteristic of the fintechs must be highlighted, mainly if we think on the context of a country such as Brazil, in the semi-periphery of the world-system. It consists in the potential that these companies present to enlarge financial inclusion mechanisms, once—at least theoretically—, for having leaner structures and for being capable of operating wherever there is a minimum infrastructure of telecommunications, both regulatory authorities and the "traditional" financial agents themselves see fintechs as an opportunity for the expansion of offer of financial services and products for small companies and people with lower income.

This optimist view is embraced also in the Interamerican Development Bank Report that advocates the idea that fintechs may provide better general efficiency for the financial system (and for the global development of digital economy), as

Table 3.7 World—Distribution of fintechs (opened between 2005 and 2015)

Rank	Country	No. of fintechs	Rank	Country	No. of fintechs	Rank	Country	No. of fintechs
1°	USA	3069	37°	Malaysia	17	73°	Myanmar	3
2°	UK	695	38°	Estonia	16	74°	Zimbabwe	3
3°	India	262	39°	Nigeria	16	75°	Azerbaijan	2
4°	Canada	247	40°	Bermuda	15	76°	Croatia	2
5°	China	160	41°	Ghana	15	77°	Dominica	2
6°	Australia	154	42°	Colombia	14	78°	Gibraltar	2
7°	Singapore	139	43°	Luxemburg	14	79°	Iran	2
8°	Germany	135	44°	Egypt	13	80°	Jordan	2
9°	France	132	45°	Norway	13	81°	Malta	2
10°	Brazil	79	46°	Ukraine	12	82°	Namibia	2
11°	Netherlands	75	47°	Vietnam	11	83°	Slovenia	2
12°	Spain	70	48°	Czech Republic	10	84°	Sri Lanka	2
13°	Israel	66	49°	Portugal	10	85°	Algeria	1
14°	Hong Kong	63	50°	Greece	9	86°	Barbados	1
15°	Sweden	63	51°	Latvia	9	87°	Belarus	1
16°	Russia	62	52°	Lebanon	9	88°	Belize	1
17°	South Africa	60	53°	Mauritius	9	89°	Botswana	1
18°	Mexico	56	54°	Thailand	9	90°	Costa Rica	1
19°	Switzerland	55	55°	Bulgaria	8	91°	Cote d'Ivoire	1
20°	Ireland	50	56°	Caymans Islands	8	92°	Dominican Rep.	1
21°	Italy	44	57°	Cyprus	8	93°	El Salvador	1
22°	Denmark	42	58°	Iceland	8	94°	Guatemala	1
23°	Japan	41	59°	Hungary	7	95°	Isle of Man	1
24°	Belgium	38	60°	Slovak Rep.	7	96°	Jersey	1
25°	Indonesia	32	61°	Romania	6	97°	Macedonia	1
26°	Chile	31	62°	Bangladesh	4	98°	Malawi	1
27°	Argentina	30	63°	Lithuania	4	99°	Morocco	1
28°	Finland	27	64°	Pakistan	4	100°	Puerto Rico	1
29°	Korea (Rep.)	26	65°	Panama	4	101°	Qatar	1
30°	Poland	26	66°	Peru	4	102°	Seychelles	1

(continued)

Table 3.7 (continued)

Rank	Country	No. of fintechs	Rank	Country	No. of fintechs	Rank	Country	No. of fintechs
31º	New Zealand	24	67º	Rwanda	4	103º	Sierra Leone	1
32º	Turkey	24	68º	Uganda	4	104º	Togo	1
33º	Austria	20	69º	Tanzania	4	105º	Trinidad y Tobago	1
34º	Philippines	20	70º	Bahrain	3	106º	Uruguay	1
35º	Kenya	19	71º	Taipei	3	107º	Zambia	1
36º	United Arab Emirates	19	72º	Guernsey	3			

Source Haddad and Hornouf (2018, p. 17)—based on Crunchbase

well as in the eradication of exclusion and promoting greater financial (and social) inclusion (BID 2017, 7). Due to the fact that in Latin America an important share of the population is not even banked (and have no access to "formal financial services"), as well as for the great financial difficulties which small and medium companies are subject (suffocated by the conditions imposed by the large banks), fintechs may be a fundamental element to suppress these demands, contributing for the economic development of countries in the region.

In the context of the 393 fintechs researched in the IDBs Report, in 18 Latin American countries, they found some "common denominators" in terms of action, which confirms the important role of fintechs in the financial inclusion process. According to the research—which is apologetic in several ways—significant part of the Latin American fintechs was destined to serve the segments until now neglected by the financial system: "41.3% of the interviewed affirmed that their mission is to assist the clients that continue to be excluded or poorly assisted by the traditional financial services sector, whether as individuals or small and medium-sized enterprises (SMEs)" (BID 2017, 17). This scope of fintechs action would be even more significant if we compare them with the traditional banks that tend to offer products and services almost exclusively for formalized companies with some degree of financial strength. In case of fintechs, it would be common to offer cheaper products and services, without formal/major bureaucratic requirements, what would allow them to assist also small firms and economic agents. Yet, the lower costs incurred in the offer of digital channels—typical of fintechs—added to the emergence of services destined for segments not assisted or poorly served, in a more accessible way, could then allow a more broad financial inclusion in the region (BID 2017, 17).

In case of demand made by small- and medium-size companies, the research of BID registered also that as follows:

1. One in every four fintech startups considers that the small- and medium-sized enterprises (SMEs) are their main client (independent of already being banked or not);
2. Their mission would be to offer them different solutions that are beyond the alternative financing tools and include also collection, digital accountancy, international payments, and financing of invoices or factoring, among others (BID 2017, 28);
3. Alternative financing models—such as the collective financing (crowdfunding)—benefit especially smaller companies and startups, as well as those that have no access to traditional credit;
4. The digital history of their transactions—or other digital fingerprints—may be used in new methodologies for risk assessment, partially mitigating the asymmetries of information that difficult assessment of credit risk (BID 2017, 28).

There is consequently, a great hope that fintechs may enlarge the access of smaller economic agents to products and financial services, allowing them an improvement in their performance, and greater proactivity in the division of labor in which they participate. The literature, as well as the statistics, on the other hand, show that fintechs will have greater difficulty in promoting an actual rupture in the financial market which are part, as evidently shows the case of digital banks: for each case of success of fintechs, the traditional companies already consolidated in the market probably will buy those fintechs that stands out, emptying their disruptive potential.

We can add to this the fact that in Latin American countries the political and economic power of banks is overdone, as well as official regulation directly favor the sector (or at least protects it, in case of some real threat, under the pretext of financial stability of the economic system). As it seems, there is not in the near time horizon a real transformation in the sector, neither a structural change or some that could at least empower financial firms that are more close—socially and physically—to the necessities of people where they actually live. In this sense, the Schumpeterian "creative destruction," latent in case of disruptive power of fintechs, probably will not occur.

References

ABFintech (2018) Brazil Fintech Deep Dive 2018. São Paulo: PWC/ABFintech, 56 pp. https://www.pwc.com.br/pt/setores-de-atividade/financeiro/2018/pub-fdd-18.pdf. Access in 13 March 2019

Abramovay R (2004) A densa vida financeira das famílias pobres. In: Abramovay R (org) Laços financeiros na luta contra a pobreza. Annablume/FAPESP, São Paulo, pp 21–67

Accorsi A (2014) O banco do futuro: perspectivas e desafios. Revista de Administração 49(1):205–216

Amado AM (2006) Impactos regionais do processo de reestruturação bancária no início dos anos 1990. In: Crocco M, Jaime Jr, FG (eds) Moeda e território: uma interpretação da dinâmica regional brasileira. Autêntica, Belo Horizonte, pp 147–168

BCB. Banco Central do Brasil (2016) Relatório de Estabilidade Financeira
BCB. Banco Central do Brasil (2017) Concentração, Competição e Inovação. Relat Econ Banc 1(1). http://www.bcb.gov.br/pec/depep/spread/REB_2017.pdf. Accessed 31 July 2018
BCB (2018) Panorama do sistema nacional de crédito cooperativo. Banco Central do Brasil, Brasília 23 p. https://www.bcb.gov.br/content/estabilidadefinanceira/coopcredpanorama/9_panorama_sncc_2018.pdf Accessed 03 June 2019
Banco Palmas (2018) Banco Palmas 20 anos conectados. Banco Palmas, Fortaleza, 7 pp. http://www.institutobancopalmas.org/manifesto-20-anos-banco-palmas/. Accessed 23 Oct 2018
BCB/SNAC (2016) Panorama do Sistema Nacional de Crédito Cooperativo. Banco Central do Brasil, Brasília
BCB/SNAC (2017) Panorama do Sistema Nacional de Crédito Cooperativo. Brasília: Banco Central do Brasil
BID. Bank of Interamerican Development (2017) Fintechs. Innovations you may not know were from Latin America and the Caribbean. BID y Finnovista. Maio de 2017. https://publications.iadb.org/handle/11319/8265?locale-attribute=pt&. Accessed 12 April 2018
Brown E, Castañeda F, Cloke J, Taylor P (2013) Towards financial geographies of the unbanked: international financial markets, 'bancarizacion' and access to financial services in Latin America. Geograph J 179(3):198–210
Burigo FL (2006) Finanças e solidariedade: o cooperativismo de crédito rural solidário no Brasil. Estud Sociol Agrícola 14(2):312–349
Canton AM (2010) A Rede Lotérica no Brasil. IPEA/CEF, Brasília
Castells M (1996) The rise of the network society. Information Age, vol I. Blackwell Publisher, Oxford
Chick V (1994) A evolução do sistema bancário e a teoria da poupança, do investimento e dos juros. Ensaios FEE 15(1):9–23
CNC (2018) Confederação Nacional do Comércio. Pesquisa CNC. Endividamento e Indadimplência do Comsumidor. 12 pp. http://cnc.org.br/sites/default/files/arquivos/perfil_de_endividamento_das_familias_brasileiras_em_2017.pdf. Accessed 12 Dec 2018
Contel FB (2011) Território e finanças. Técnicas, Normas e Topologias Bancárias no Brasil. Annablume, São Paulo
Crocco M, Santos F, Figueiredo A (2013) Exclusão financeira no Brasil: uma análise regional exploratória. Rev Econ Política 33(3):505–526
de França Filho GC, Silva JT Jr (2009) Bancos Comunitários de desenvolvimento. In: Cattani AD, Laville JL, Gaiger LI, Hespanha P (eds) Dicionário Internacional de Outra Economia. Coimbra, Almedina, pp 31–36
De Paula CG (2015) Do Território ao lugar: bancos comunitários, moedas locais e o circuito inferior da economia urbana em São Paulo (SP). Departamento de Geografia/Universidade de São Paulo (Unpublished Master Degree Dissertation), São Paulo
De Paula CG (2018) Bancos comunitários e moedas sociais: racionalidades e contra-racionalidades financeiras no território. Departamento de Geografia/Universidade de São Paulo (Unpublished Qualification examination report), São Paulo
Demirguc-Kunt A, Klapper L, Singer D, Van Oudheusden P (2015) The Global Findex Database 2014. Measuring financial inclusion around the world. World Bank, Washington, 97 pp
Dos Santos WG (2017) A democracia impedida. O Brasil no Século XXI. FGV Editora, Rio de Janeiro
Dow S (1999) The stages of banking development and the spatial evolution of financial systems. In: Martin R (ed) Money and the space economy. Wiley, New York, pp 31–48
Dymski G (2005) Banking strategy and financial exclusion: tracing the pathways of globalization. Economia (Curitiba) 31(29):107–143
Dymski G (2007) Exclusão e eficiência: a transformação global do core banking, um estudo de caso sobre o Brasil. In: De Paula LF, Oreiro JL (eds) Sistema Financeiro. Uma análise do setor bancário brasileiro. Elsevier, Rio de Janeiro, pp 255–284

Barone, F. e Sader E (2008) Acesso ao crédito no Brasil: evolução e perspectivas. Revista de Administração Pública—FGV 42(6):1249–1267
FEBRABAN. Federação Brasileira de Bancos (2011) Bancarização e Inclusão Financeira no Brasil. FEBRABAN/FELABAN, São Paulo, 50 pp
Feltrin LE, Cruvinel E, Dodl A (2009) Inclusão financeira no Brasil. Projeto estratégico do Banco Central. In: Feltrin LE, Cruvinel E, Dodl A (eds) Perspectivas e desafios para a inclusão financeira no Brasil: visão de diferentes atores. Banco Central do Brasil, Brasília, pp 13–26
FINTECHLAB. (2017). FintechLab Report. Clay Innovation. 48 pp
Freire MV (2013) A importância dos bancos comunitários para a inclusão financeira. In: Instituto Palmas/NESOL-USP (orgs) Banco Palmas. 15 Anos. A9 Editora, São Paulo, pp 41–59
Global Findex. Global Financial Inclusion Database. (several years). New York: World Data Bank. https://globalfindex.worldbank.org/
Haddad C, Hornuf L (2018), The emergence of the global fintech market: economic and technological determinants. Small Business Economics (On-line) 2018. https://doi.org/10.1007/s11187-018-9991-x
Hendrikse R, Bassens D, van Meeteren M (2018) The appleization of finance: charting incumbent finance's embrace of FinTech. Finance Soc 4(2):159–180
Hepworth M (1990) Geography of the information economy. Guilford Press, London
Kraychete E (2004) Entre o microcrédito e a 'bancarização': impasses para a constituição de um mercado de microfinanças adequado às necessidades dos microempreendedores. In: IX Congreso Internacional del CLAD sobre la Reforma del Estado y de la Administración Pública, Madrid, España, 2–5 Nov
Kumar A (ed) (2004) Brasil: acesso a serviços financeiros. Banco Mundial/IPEA, Rio de Janeiro
Law J (ed) (2008) Oxford dictionary of finance and banking. Oxford University Press, Oxford
Lee R (1999) Local Money: geographies of resistance and autonomy? In: Martin R (ed) Money and the space economy. Wiley, New York, pp 207–223
Lagoarde-Segot T (2015) As finanças solidárias. Um humanismo econômico. Edições Piaget, Lisboa
Levorato W (2009) Microfinanças e o sistema bancário brasileiro. In: Feltrin LE, Cruvinel E, Dodl A (eds) Perspectivas e desafios para a inclusão financeira no Brasil: visão de diferentes atores. Banco Central do Brasil, Brasília, pp 169–187
Leyshon A (1995a) Geographies of money and finance I. Progr Human Geogr 19(4):531–543
Leyshon A (1995b) Annihilating space? The speed-up of communications. In: Allen J, Hamnett C (eds) A shrinking world? Oxford University Press, Oxford, pp 11–54
Lojkine J (1992) La Révolution informationnelle. PUF, Paris
Menezes MS, Crocco MA (2009) Sistemas de moeda local: uma investigação sobre seus potenciais a partir do caso do Banco Bem em Vitória/ES. Econ Soc 18(36):371–398
Montenegro MR (2010) O circuito inferior no centro de São Paulo frente às dinâmicas da globalização e ao uso corporativo do território. In: Revista Tamoios (Impresso), Ano IV, pp 29–41
Montenegro MR (2014) Globalização, trabalho e pobreza nas metrópoles brasileiras. FAPESP/Annablume, São Paulo
Neiva AC, Braz JOB, Tsukumo DJ, Melo J (2013a) Solidarity finance and public policy: the brazilian experience of community development banks. In: The United Nations non-governmental Liaison service (NGLS). Working Paper 19 pp. http://www.unrisd.org/80256B42004CCC77/(httpInfoFiles)/07496B7B589AF099C1257B7400317C32/$file/Neiva%20et%20al.pdf. Accessed 23 June 2018
Neiva AC, Braz JOB, Nakagawa CT, Mascarenhas TS (2013b) Banco Palmas: resultados para o desenvolvimento comunitário e a inclusão financeira e bancária. In: Instituto Palmas/NESOL-USP (orgs.). Banco Palmas. 15 Anos. A9 Editora, São Paulo, pp 105–178
North PJ (2005) Scaling alternative economic practices? Some lessons from alternative currencies. Trans Inst Br Geogr 30(2):221–233

North PJ (2017) Geographies of alternative, complementary and community currencies. In: Pollard J, Martin R (eds) Handbook on the geographies of money and finance. Edward Elgar, Cheltenham, pp 518–537

Palhares VMA (2004) Análise histórica e evolutiva do cooperativismo de crédito no Brasil. In: Pinho DB, Palhares VMA (eds.) O Cooperativismo de Crédito no Brasil. Do Século XX ao Século XXI. Esetec Editores Associados, Santo André, pp 33–76

Parserisas DD (2012) Circuitos de la economía urbana y sistema financiero de crédito. Un análisis en la ciudad de Olavarría, Provincia de Buenos Aires, Argentina. Editorial Académica Española, Saarbrücken

Philippon T (2017) The Fintech opportunity. BIS Working Papers No. 655. Bank For International Settlements, Basel, 33 pp

Pike A, Pollard J (2010) Economic geographies of financialization. Econ Geogr 86(1):29–51

Pinho DB (2004) Brasil. Sistemas de crédito cooperativo pioneiro, sindical e solidário. In: Pinho DB, Palhares VMA (eds) O Cooperativismo de Crédito no Brasil. Do Século XX ao Século XXI. Esetec Editores Associados, Santo André, pp 11–32

Rankin KN (2013) A critical geography of poverty finance. Third World Q 34(4):547–568

Rigo AS, França Filho GC (2017) Bancos Comunitários e moedas sociais no Brasil reflexões a partir da noção de economia substantiva. Edufba (Ebook), Salvador. https://repositorio.ufba.br/ri/handle/ri/24260. Accessed 03 Feb 2019

Rojas-Suárez L (2007) Promoviendo el Acceso a los Servicios Financieros ¿Qué nos Dicen los Datos sobre Bancarización en América Latina? Federación Latinoamericana de Bancos/FELABAN, 38 pp

Santiago EG (2003) Cadeias produtivas locais. In: Segundo JJMN, Magalhães S (eds) Bairros pobres, ricas soluções. Banco Palmas ponto a ponto. Expressão Gráfica, Fortaleza, p 31/32

Santos M (1994a) Técnica, Espaço, Tempo. Globalização e meio técnico-científico informacional. Hucitec, São Paulo

Santos M (1994b) Por uma Economia Política da Cidade. O caso de São Paulo. Hucitec, São Paulo

Santos M ([1975] 1979) The Shared Space: the two circuits of the urban economy in underdeveloped countries. Methuen, London

Santos MP, Rodrigues J, Medina G (2017) Cooperativismo em Goiás: como equalizar competitividade e solidariedade? Interações 18(4):31–42

Segundo JJMN, Magalhães S (2003) Um banco diferente. In: Segundo JJMN, Magalhães S (eds) Bairros pobres, ricas soluções. Banco Palmas ponto a ponto. Expressão Gráfica, Fortaleza, pp16–19

Silva SP (2017) Economia solidária e finanças de proximidade: realidade social e principais características dos empreendimentos de finanças solidárias no Brasil. In: IPEA. Textos para discussão no. 2270, 75 pp

Silveira ML (2007) Metrópolis brasileñas: un análisis de los circuitos de la economía urbana. Eure 33(100):149–164

Silveira ML (ed) (2016) Circuito de la economía urbana. Ensayos sobre Buenos Aires y São Paulo. Editorial Café de las Ciudades, Buenos Aires

Silveira ML (2017) Banalidade das finanças e cidadania incompleta: lugar e cotidiano na globalização. In: Geousp—Espaço e Tempo (Online) 21(2):370–383

Singer P (2002) Introdução à Economia Solidária. Fundação Perseu Abramo, São Paulo

Singer P (2009) Finanças solidárias e moeda social. In: Feltrin LE, Cruvinel E, Dodl A (eds) Perspectivas e desafios para a inclusão financeira no Brasil: visão de diferentes atores. Banco Central do Brasil, Brasília, pp 67–78

Singer P (2013) O Banco comunitário de desenvolvimento como política pública de economia solidária. In: Instituto Palmas/NESOL-USP (orgs) Banco Palmas. 15 Anos. A9 Editora, São Paulo, pp 35–39

Soares MM, Sobrinho ADM (2008) Microfinanças. O Papel do Banco Central do Brasil e a Importância do Cooperativismo de Crédito. Banco Central do Brasil, Brasília, 184 pp. https://www.bcb.gov.br/htms/public/microcredito/livro_microfinan%C3%A7as_internet.pdf

Thrift N, Leyshon A (1997) Spatial financial flows and future global geography. In: Mabogunje A (ed) The state of the earth. Blackwell, Oxford, pp 279–304

Tickell A (2000) Finance and localities. In: Clark G, Feldman M, Gertler B (eds) The Oxford handbook of economic geography. Oxford University Press, Oxford, pp 230–247

Toscano I (2003) História e lutas: da remoção ao bairro Conjunto Palmeira. In: Segundo JJMN, Magalhães S (eds) Bairros pobres, ricas soluções. Banco Palmas ponto a ponto. Expressão Gráfica, Fortaleza, pp 7–15

Tsai C-H, Peng K-J (2016) The fintech revolution and financial regulation: the case of online supply chain financing. Social Science Research Network (SSRN Papers), 32 pp. https://papers.ssrn.com/sol3/papers.cfm?abstract_id=3035386. Accessed 16 Aug 2018

VALOR ECONÔMICO (2018) Mais de 400 fintechs disputam mercado. Carmen Nery para Valor Econômico. Disponível em https://www.valor.com.br/empresas/4334666/mais-de-400-fintechs-disputam-mercado. Accessed on 19 July 2018

Wójcik D, Cojoianu T (2018) A global overview from a geographical perspective. In: Cassis Y, Wójcik D (eds) International financial centres after the global financial crises and Brexit. Oxford University Press, Oxford, pp 207–231

Yunus M (2001) Hacia un mundo sin pobreza. Editorial Andres Bello, Barcelona/Buenos Aires

Conclusions

Our study sought to outline the main characteristics of the financialization process of the Brazilian territory, seeking to work with a definition of financialization that would draw attention to the action of all financial agents, highlighting commercial banks and the spread of credit operations in the national space (the basis of low-intensive financialization). We identified this phenomenon as from three levels—or scales—of determination: the global scale (that relates to the contemporary capitalism with financial dominance), the national scale (the Brazilian sociospatial formation), and some phenomena in regional and local scale, which allowed to clarify the specificities of financialization processes that are closer to the everyday life of small economic agents and ordinary people living in the Brazilian territory.

It is precisely in the analysis of this dialectic between the movement of the world (the historical totalization process) and the particular events (the concretization of the movement of the world in the territories, regions, and places) that we can recognize both the logic of the geographical space organization and its possible trends of reorganization. As proposed by the French philosopher Sartre (1960), the historical totalization process does not take place without the materiality that characterizes social life on the most different scales, the materiality that he named "practico-inert." And it is justly this labor and technique-intensive material objects that mainly condition contemporary human action and division of labor. The dialectic between objects and actions, by the way, defines geographical space, as already shown by one of the epigraphies of this book (Santos 1994b) and as we argued on several parts of this work, to better understand the geographic features of financial actors and processes.

In national scale, it was also possible to acknowledge how the regional organization of the Brazilian banking system was transformed into a more centralized one, having São Paulo as the main financial center of the Brazilian urban network. Because of the dimension of the networks controlled by the headquarters of large private banks and financial firms located in São Paulo, and also due to the fact that the only Stock Exchange in Brazil is placed there as well, it is correct to say that the Metropolitan Region of São Paulo is the core of this vast financial-informational hinterland that spreads through the Brazilian urban network.

F. B. Contel, *The Financialization of the Brazilian Territory*, Economic Geography,
https://doi.org/10.1007/978-3-030-40293-8

The book had, in this sense, two main ambitions: one of more empiric nature, to show with data, statistics, official laws, and historical facts how the Brazilian financial and banking system developed, under a geographic vision. But, it also had a more theoretical—or abstract—concern, seeking to identify forms of interpretation of financialization of geographical space that often do not appear explicitly in the recent literature about the theme.

The first form of interpretation which we tried to consider in our study was the effective incorporation of technical systems in the analysis of financialization of geographic space, in its most different scales and forms of existence. Since the macro-technical systems represented by the submarine cables, telephony, and artificial satellites networks, up to the most miniaturized technical objects such as smartphones, points of sales (terminals), or credit and debit cards, the paths of spatial diffusion of finances are in great part determined by the (pre)existence of these support-networks and technical systems that allow their circulation, increasingly expansive, fluid and fast. Technical systems, therefore, are an important part of determining financial processes, along with elements that are more "purely" economic, as well as political, cultural, and social ones. These technical systems are responsible for the installation of financial agents and their products in virtually all places and regions of the world, reaching almost all populations and social classes, which constitutes a process that could be denominated "credit hypercapillarity" (Contel 2011). It was only with this new technical basis that globalized finances became almost ubiquitous, albeit in an extremely selective and unequal feature, giving concrete existence to the phenomenon identified by Buhkarin (1915) as the intrinsic *pervasiveness* of finance in the early twentieth century. Although some authors—mainly geographers—have been concerned with the issue of technical systems that are on the basis of the action of financial agents, in many texts and important discussions this fact remains as a "hidden dimension" and undervalued in these analysis.

In the set of different technical systems that are on the basis of the current globalization phenomenon, attention shall also be driven to the importance of information techniques as central element of this rise of importance of finances in the contemporary world. Finances and information, as shown by different authors such as Goldfinger (1986), O'Brien (1992), Santos (1994a), Joel Kurtzman (1994), and Porteous (1995), are two elements that are increasingly conjugated, and this symbiosis constitutes one of the main characteristics of contemporary financialization process. This symbiosis between money and information is seen in first place in the dematerialization process of money; it is increasingly common, even in countries of the periphery of the capitalist system, the reduced daily use of bills and coins, for the use of digital means of payment, which allow the execution of commercial transactions as from electronic currencies. Not only in this "lower circuit" of the national financial systems this combination of finances/information is identifiable, as seen in the most anti-establishment form of finance organization, the community banks (item 3.3). The entire recent process of securitization of economies, with the emergence of new financial products (such as derivatives, swaps, options, among others), is also a manifestation of this "computerization of finance" in a high scale. Finally, as shown in item 3.4 of this work, fintechs are the clearest example of the interrelationship between information technology and finance, opening new fields for operation of agents from this sector, and also for transforming the financial system as a whole. With fintechs, information-intensive technologies have definitely taken the place of previously existing material networks, becoming part of the daily work of such firm's elements and processes as artificial intelligence, data analytics, cloud services, distributed ledger technologies, biometric and identity management, cybersecurity, robot advisors, among so many other novelties that tend to significantly break the whole recent pattern of financialization of geographic space.

The last central element of our discussion refers to legal norms and standards. The norms, especially the official/legal ones, are a fundamental aspect to understand this primacy that finances gained in the contemporary world, as well as to understand the logics of the operation of the financial system. As we have mentioned in Chap. 2, laws and norms may be considered as a constitutive element of the geographic space itself, alongside with technical, economic, political, and social contents. The Nation state legislations that concern territories are the most well-finished form of these normative contents of geographic space. The fundamental laws of this system, incidentally, are, in general, the main beacons to define the periodization of the financialization of territory, considering the power that they have to regulate financial processes and actors in a national space. The nature of these norms and laws, on the other hand, have been a powerful element for naturalizing and legitimizing practices and nexuses that are more related to the needs of financial firms themselves, then with the needs of all other economic agents—and the civil society in general. Interfere in these normative contents is indeed a fundamental and *sine qua non* condition to establish a more democratic and popular use of financial resources worldwide.

As much as we are living in a historical period in which finances are much more determinant for the reproduction of populations in their everyday life, for the performance of financial and non-financial companies, as well as for the power of national States, finances do not have an absolute autonomy in respect to other social instances, including the instance which is the geographic space. Although more movable, ubiquitous, and rapid, financial circulation depends on support-networks for their fluidity—especially in long distances—as well as are dependent of technical systems and objects that allow the economic agents to manipulate these resources (data centers, personal computers, smartphones, points of sales, credit and debit cards, etc.). Albeit more powerful than the other economic circuits (agriculture, industry, commerce, and other services), finances are always dependent on the public authorities that regulate them, as multilateral organizations and the nation States themselves, which authorize or not certain processes and behaviors of financial agents. Although they are also preponderant in the contemporary economy, finances always need to make a "return to the real economy" for their valorization, mainly in moments of large financial crises, when the frenetic pace of self-valorization as "fictitious capital" is obstructed and socially delegitimized. Possibly the most thorough manifestation that finances are not an autonomous element of organization of contemporary economy is the proper formation of international—and national—financial centers that operate as large investment and recycling financial sites and constitute poles of attraction that finance cannot escape or avoid (especially "high finance"). As we have also tried to demonstrate, the processes of financial inclusion and exclusion are also manifestations of this dependence on the finances of other elements of the geographical space.

In this context of greater "relative autonomy" of finances, it should be finally pointed out that are emerging recently alternative forms to use finances, within this hyperfinancialized capitalism. This process also reveals that, however, powerful or invasive the agent or process in question, its effective repercussion in places is not unequivocal, and may provoke unanticipated—and even counter-rational—local dynamisms to the original logics of these agents.

In the Brazilian territory, some new forms to use finances indicate that other logics may guide the contemporary financialization of geographic space. Solidary and alternative uses, which value the cooperation and the local trust networks, have become increasingly common, such as the case of credit cooperatives, analyzed in this book. More than cooperatives, the community banks are also a clear expression of alternative forms of finance organization, beyond the capitalist instrumentality that currently is the basis of financial operations. These alternative paths are very close to what Santos (1996) called the creation of "horizontalities" (*horizontalidades*) in places and regions, that is, the production of new logics of space organization that takes advantage of the external vectors of modernity to reconfigure them, producing a built environment and everyday life that are more just and solidary, aiming at the economic and political emancipation of these local or regional collectivities.

The fintechs themselves, even if they are firms that emerged in the most acute financialization in which we live, bring with them the germ of the structural transformation of financial systems that they are part, considering the disruptive capacity that these startups have in respect to the "business models" of hegemonic financial companies. Analyzing these alternative financial agents may be one of the most propitious and suitable ways that the academic studies may provide for the structural transformation of the contemporary financial system, transformation that could, in its turn, gradually move us toward a socially, economically, and geographically more just and egalitarian world.

References

Bernardes A (2002) São Paulo, produção de informações e reorganização do território brasileiro. Departamento de Geografia/Universidade de São Paulo (Unpublished PhD Thesis), São Paulo
Bukharin N (1915) Imperialism and world economy. Martin Lawrence Limited, London
Contel FB (2009) Espaço Geográfico, Sistema Bancário e a Hipercapilaridade Do Crédito No Brasil. Caderno CRH 22(55):119–134
Contel FB (2011) Território e finanças. Técnicas, Normas e Topologias Bancárias no Brasil. Annablume, São Paulo
Dörry S (2017) Regulatory spaces in global finance. In: Martin R, Pollard J (eds) Handbook on the geographies of money and finance. Edward Elgar, Cheltenham, pp 415–433
Goldfinger C (1986) La Géofinance. Pour compreendre la Mutation Financière. Seuil, Paris
Katrin K (2013) A critical geography of poverty finance. Third World Q 34(4):547–568
Kurtzman J (1994) The death of money. How the electronic economy has destabilized the world's markets and created financial chaos. Simon & Schuster, New York
Monbeig P (1957) Capital e geografia. In: Monbeig P (ed) Novos Estudos de Geografia Humana Brasileira. Difel, São Paulo
O'Brien R (1992) Global financial integration. The end of geography. The Royal Institute for International Affairs, New York
Porteous D (1995) The geography of finance. Spatial dimensions of intermediary behaviour. Ashgate Publishing Limited, Aldershot
Santos M (1994a) Técnica, Espaço, Tempo. Globalização e meio técnico-científico informacional. Hucitec, São Paulo
Santos M (1994b) Por uma Economia Política da Cidade. O caso de São Paulo. Hucitec, São Paulo
Santos M (1996) A Natureza do Espaço. Técnica e Tempo, Razão e Emoção. São Paulo: Hucitec. (Translated to french in *La Nature de L'Espace*, Paris: L'Harmattan, 1997)
Santos M (1999) O país distorcido. Publifolha, São Paulo
Santos M (2017) Toward an other globalization: from the single thought to universal conscience. Springer, Heidelberg. Original work published in 2000
Sartre JP (1960) Critique de la raison dialectique. Gallimard, Paris
Yunus M (1997) Hacía un mundo sin pobreza. Editorial Andres Bello, Santiago de Chile

F. B. Contel, *The Financialization of the Brazilian Territory*, Economic Geography,
https://doi.org/10.1007/978-3-030-40293-8

MIX
Papier aus verantwortungsvollen Quellen
Paper from responsible sources
FSC® C105338

FSC
www.fsc.org

If you have any concerns about our products,
you can contact us on
ProductSafety@springernature.com

In case Publisher is established outside the EU,
the EU authorized representative is:
Springer Nature Customer Service Center GmbH
Europaplatz 3, 69115 Heidelberg, Germany

Printed by Libri Plureos GmbH
in Hamburg, Germany